THE CFZ YEARBOOK

1998

Edited by
Jonathan Downes
and
Graham Inglis

Typeset by Jonathan Downes,
Cover and Layout by die katze orangebaum for CFZ Communications
Using Microsoft Word 2000, Microsoft , Publisher 2000, Adobe Photoshop CS.

Photographs © 2008 CFZ except where noted

First published in Great Britain by CFZ Press

CFZ Press
Myrtle Cottage
Woolsery
Bideford
North Devon
EX39 5QR

© CFZ MMVIII

ISBN: 978-1-905723-27-0

Contents

INTRODUCTION TO THE 2008 EDITION

I am finding the ongoing programme to reassure the first eight CFZ Yearbooks in improved perfect bound editions both a rewarding, and an oddly frustrating process. Back when the first three Yearbooks were first published, I made sure that all the masters were available. I had learnt my lesson back in the early 1990s when I was working for a well-known pop singer who was anxious to reissue his mid 1970s LPs in the then new CD format. I spent many happy, though frustrating, hours wading my way through the catalogue of the EMI vaults to see what recordings were available. To my frustration then I found that many key variations of songs which I wanted to include on CD reissues were missing, and that I would have to make a series of compromises in order to complete the product that I wanted to release.

So, as - right from the beginning - I had always intended that the CFZ Yearbooks would provide a library of cryptozoological information for many years to come, I kept the masters. When, over a decade later, I came to start work on this present project, I found - much to my annoyance and frustration - that many of them had disappeared. Oll Lewis and I, therefore, have spent some months piecing together the best quality versions that we can of the 1997 and 1998 Yearbooks. The quality of neither volume is as good as we would have liked, but finally the 1998 volume is finished.

That has dealt with my frustration. Why, you may well ask, have I found the process rewarding? Well that is a simple question to answer. I had simply forgotten quite how good the 1998 volume was.

1997 was a strange year for me personally. It was the year when, following the unprecedented upsurge of interest in things fortean, mostly fuelled by the 50th anniversary of the so-called Roswell Incident, that I suddenly found myself in the public eye having achieved a modicum of fame. For the first time in my adult life I was earning quite a good wage, and was able to follow my cryptoinvestigative instincts, and carry out a series of UK based cryptozoological investigations. It was also the only year in my adult life when I lived by myself. A year before I had been married, and the year later my old mucker Richard Freeman moved in to live with me. But in 1997 I was living a bachelor life with my dog Toby, and - together with Graham Inglis - was beginning to teach myself how to run a successful publishing company.

As even a cursory look at these pages will tell, we were still at a very low part of the learning curve, and my typesetting skills left an awful lot to be desired. In my defence, however, the Yearbooks were my first attempts at a long form publication, and at that time we didn't even have a PC. The whole thing was put together with scissors and glue, and a rudimentary page setting programme on an Amiga games computer.

I decided to republish the 96,97, and 98 Yearbooks in facsimile, rather than to have them re-typed, because I feel that they are important historical documents as regards the history of the CFZ. I abhor the currently accepted practice of going back and rewriting your own history as soon as you are in the position to do so. The CFZ started from humble beginnings, and has gone from those beginnings to achieve some fairly spectacular things, and will - I sincerely hope - go on to further heights of achievement in years to come. I was proud of the 1998 Yearbook when it first came out, and although - looking back - there are things that we could have done better, it was a pretty good achievement in those halcyon days before the Pentium chip, broadband, and the universiality of the Internet changed the world forever.

Read and enjoy. I know I have.

Jonathan Downes,
Centre for Fortean Zoology,
North Devon,
April 2008

INTRODUCTION

Dear Friends,

Welcome to the 1998 Yearbook. Despite our brave attempts to get it finished in time for Christmas 1997 we are once again running a little late, but we hope that you think that the final result has been worth it. The content this year is as varied and complex as ever, but the main themes are vaguely aquatic with articles from Richard Freeman on giant crocodiles, from Chris Moiser on the River God of the Zambezi, and a mammoth exploration of monster-haunted lakes by Michael Playfair.

We also have an interesting addendum to last year's lengthy listing of films with a cryptozoological theme from Neil Arnold who has uncovered a veritable treasure trove of obscure movies which were not included in last year's listing (also by Mike Playfair).

We hope that you enjoy this year's collection and that you will continue to support us throughout 1998.

Best wishes,

Jon Downes and Graham Inglis

Giant Crocodiles -
The Ultimate Predators

by Richard Freeman

Man has a habit of exaggeration; that is how many 'monsters' are created. The more dangerous (in our perception) the animal, the greater the tales we weave around it. One only has to look at the film world: *Jaws, Grizzly, Alligator* - and, more recently, the preposterous *Anaconda*, a film as much in touch with reality as Erich von Daniken!

Few carnivorous animals are anything like as dangerous to man as some of us would like. It is perhaps fitting that the one type of animal that has the most persistent and persuasive history of 'gigantism' is also one of the few true man-eaters - the crocodile.

To be fair, only two of the 22 existing species of crocodile regularly attack humans. However, these two, *Crocodylus niloticus* (the Nile crocodile) and *Crocodylus porosus* (the Indo-Pacific crocodile account for over 5 000 annual human deaths - far more than any shark, big cat or bear. Of vertebrates, only humans themselves and the multitude of venomous snakes account for more. We must also remember that venomous snakes kill mainly in self-defence; they do not eat humans, as crocodiles do.

How does one define 'giant'? This is debatable but for argument's sake I am classing any specimen of 7 m (23 ft) or more as 'giant'. As we shall see, there are reports that suggest 23 ft specimens may look like dwarfs in comparison to the 'true giants'.

There has been a precedent for giant crocodiles in the past. *Deinosuchus hatcheri* was a giant crocodile of the late Cretaceous, flourishing in both eastern and western North America. Known only from its massive skull of 1.6 m (6 ft) long, total size estimates vary from 12 to 15 m (40 to 50 ft). This awesome animal would have been as formidable a predator as *Tyrannosaurus rex* (if somewhat less active). Indeed, some evidence suggests that, in some areas, Deinosuchus's chief prey was Albertersaurus, a smaller relative of *T. rex.*

These nightmarish brutes were not confined to the Mesozoic era. In the Tertiary a giant alligator, *Purussaurus brasiliensis*, flourished in the Amazon and rivalled Deinosuchus in size. This was only eight million years ago - long after the extinction of non-avian dinosaurs.

In the past century or so travellers have brought back tails of monster crocodiles still lingering in ill-explored corners of the tropics. It is my belief that these are not survivors of prehistoric species but are vast specimens of known species that far exceed the limits dreamed of by most zoologists.

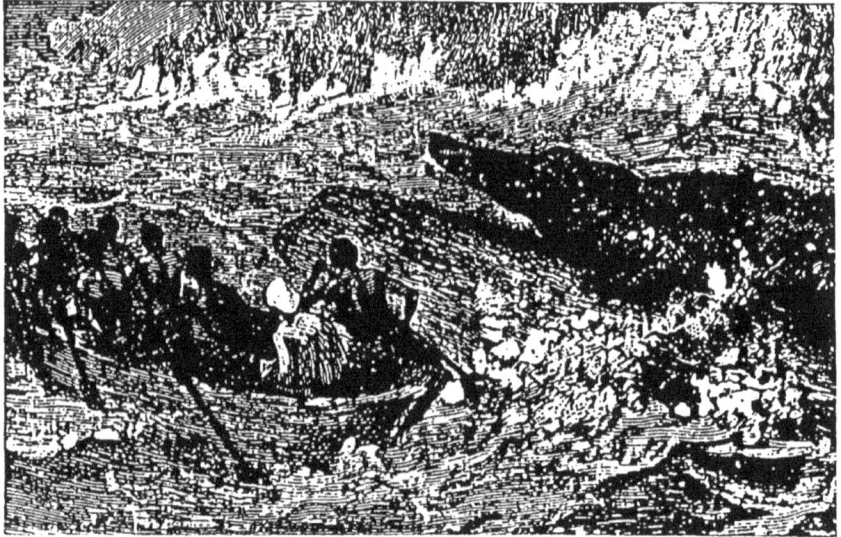

In this article I will examine the evidence for giant crocodiles and try to suggest what makes some specimens grow so frighteningly huge.

The Old World seems to have the monopoly on giant crocodilians - two New World crocodiles accorded lengths of 7 m by some authorities are *Crocodylus acutus* (the American crocodile) of Central America and the tip of Florida, and *Crocodylus intermedius* (the Orinoco crocodile) of Colombia and Venezuela. However, little if any details of such large specimens are ever given. The explorer Humbolt shot at a 6.7 m (22 ft) Orinoco crocodile in Venezuela, or so he claimed... But we only have his word for it. The maximum length for both species seems to be around 5 m (16 ft). The claims of giant crocodiles in the Old World seem much more believable, due to the sheer amount of them and the fact that some were measured my recognised experts.

The world's largest 'official' reptile is the *Crocodylus porosus* (Indo-Pacific crocodile), known also as the salt water or esuarine crocodile. However it is confined to neither salt water nor to estuaries, hence my preference for a name that indicates its geographical distribution.

The largest specimen accepted by experts was an 8.64 m (28 ft 4 in) specimen shot on the

MacArthur Bank of the Norman River, Queensland, Australia, in 1957 by Mrs Kris Pawloski. The mammoth body was too big to move but was photographed. Sadly, the photo was lost in 1968. However, Mr Ron Pawloski, a recognised authority on crocodiles, had carefully measured the specimen. He was astonished by its size, having previously measured 10 287 specimens and found none larger than 5.5 m (18 ft). It was never weighed, as it could not be moved from the tidal beach. Conservative estimates put it at around 2 tons, others at 3 tons or more.

There have, however, been some cases of exaggeration with this species. British herpetologist Angus Ballairs worked out that the head-body ratio of a typical crocodile (as opposed to specialised species like the Gharial) at 1:7.5. This had connotations for the remains of some supposed 'giants'. The skull from an '8.8 m' (29 ft) specimen killed after a 6-hour struggle in the Phillipines in 1823, by Paul de la Gironigre and George Russell, was measured at 26 inches. This gave an overall size at just under 6 m (20 ft). Another infamous giant cut down to size by Bellaids was a claimed 10 m (33 ft) crocodile harpooned in the bay of Bengal in 1940. The skull was later measured at 28 inches and the reptile's length estimated at around 6.4 m (21 ft). It then transpired that the wrong skull had been measured - the Bengal skull was smaller, meaning its owner was not even 6.4 m, let alone 10!

Persistent reports argue that - exaggeration aside - Indo-Pacific crocodiles can and do reach massive sizes.

The most famous of these is the case witnessed in the 1950s by rubber plantation owner James Montgomery. Montgomery's plantation was near the Segma River, North Borneo. He claimed to have shot 20 specimens between 6 and 8 m (20 - 26 ft) to ensure the safety of workers who washed their laundry at the river. One particular crocodile dwarfed even these. The local Seluke tribe believed it was 'the father of the devil' and threw silver coins into the water to appease it. Investigating, Montgomery found the creature on a sand bank. The crocodile filled the whole bank and had the end of its tail in the water. Wisely deciding to leave the monster well alone, he retreated. Returning later, he measured the sand bank at 9 m (30 ft), indicating that the crocodile was 10 m (33 ft) or more in length.

Today, above the Lumpar River (also in northern Borneo), there exists a similar venerated reptile. Bujang Senang - the king of crocodiles - is said to be 7.6 m (25 ft) long, and is a known man-eater, accounting for many victims. The Ibad tribe worship this lethiathan, a situation worthy of Edgar Rice Burroughs or even Conan Doyle! Another giant alive today lives in the Bhitarkanika wildlife sanctuary in Orissa state, Eastern India. It is over 7 m (23 ft) long. Three others in the sanctuary are over 6 m long.

The largest specimens - of truly mind-boggling size - have been met not in rivers but in the open sea. The Indo-Pacific has been encountered hundreds of miles from land. Larger than the biggest predatory shark (the Great White, at a maximum of 7 m) it has nothing to fear except Man.

One such sea-going encounter took place in the Gulf of Bengal in 1860. The crew and passengers on the ship *Nemesis* observed a giant crocodile at close range. One was the writer W H Marshall, who described it in his book *Four Years in Burma*:

"As the *Nemesis* was proceeding onwards towards our destination our attention was directed to an alligator of enormous length, which was swimming along against the tide (here very strong), at a rate which was perfectly astonishing. I never beheld such a monster. It passed within a very short distance, its head and half its body out of the water. I think that it could not have been less than five and forty feet long, measured from the head to the extremity of the tail and I am confident it was travelling at a rate of at least 30 miles an hour."

It should be noted that this animal would be an Indo-Pacific crocodile, not an alligator, as alligators occur in China and North America. However, early white colonials often used incorrect names for animals and these have stuck. In Australia the Indo-Pacific crocodile is often referred to as an alligator - there is even an Alligator River. In Belize, jaguars are called 'tigers' and spider monkeys 'baboons'.

Oddly, most other reports of giant crocodile-like creatures in the open sea come from well outside the Indo-Pacific crocodile's range. *The Sacramento* reported one in 1877 from the North Atlantic and, in World War 1, the German submarine U28 encountered one off the French Atlantic coast and U108 had a sighting in the North Sea. This is hardly crocodile territory. However, it should be remembered that these huge creatures would make effective gigantotherms.

Dermochelys coriacea, the leather back turtle, often strays into British waters, and some even further north. This species can weigh 2120 lbs, falling short of the largest weighed crocodile. This turtle is a gigantotherm, so large crocodiles could, in theory, enter cool waters.

The North Sea is so outside the Indo-Pacific crocodile's range, however, that such an occurrence would be like finding a giant panda in the Amazon! Perhaps these cold water crocs are a genuine totally marine undiscovered species that have evolved a 'live-bearing' system like some prehistoric reptiles. Alternatively they could be a zoomorphic phenomenon: the embodiment of our fears of the sea, given form in the shape of a giant version of the world's most dangerous predator, the crocodile.

It is fitting indeed that the other species of crocodile with a claim to monstrous proportions inhabit the Dark Continent - Africa - as this cradle of Man has a deep hold on our subconscious. If giant reptiles could lurk anywhere, it must be here. Despite the depredations of the white man and homespun warfare and famine, the heart of this awesome continent still remains an enigma.

Crocodylus Niloticus (the Nile crocodile) is the world's second-largest reptile and has long been known as a man-eater. Worshipped by the Egyptians as 'Sebek' God of the Nile, this is indeed a frightening animal. The largest 'official' specimen was shot in 1905 at Mwanza, 110 km east of Emin Pasha Gulf, by the Duke of Mecklenburg. It measured 6.60 m (21 ft).

The renowned wildlife photographer Cherry Kearton and his friend James Barns observed an 8.2m (27 ft) crocodile basking on a sandbank in the Semlik River. The size was estimated against other crocodiles and nearby objects. A photograph, which I have not seen, was published in one of Kearton's books* and apparently the crocodile in question dwarfs its companions.

A 7.9 m (26 ft) specimen was claimed by a Captain Riddick, shot at Lake Kioga. Another of similar size was killed in 1903 on the Mbaka River, which enters Lake Nyasa. This was recorded by the experienced field naturalist Hans Besser. At first he mistook the reptile for a huge canoe half drawn out of the water. It was 7.6 m (24 ft) long but had part of its tail missing! (Bitten off by a larger crocodile?!) The body was 93 cm (3.1 ft) high and was 4.6 m (14.7 ft) in girth. The skull was 1.4 m (4.48 ft) long.

The island of Madagascar has often been touted as a putative lair of giant crocodiles. These were given a name *(Crocodylus robustus)*. However these are sub-fossil remains and seem to be a giant form of the Nile Crocodile which did indeed reach a length of thirty three feet (10m). It is probable that this sub-species grew to such large proportions in order to feed on *Aepyornis maximus*, The Madagascan Elephant Bird. This giant ratite was, in terms of bulk at least, the world's largest bird and was forced into extinction by the predations of hunting humans and climatic changes. Without this large prey species the giant Madagascan crocodile became extinct leaving behind only its more modestly proportioned relatives.

The largest reported crocodiles on the African continent hail from that last great African frontier, the Congo rainforest, known to the Lingala and other Congolese people as "Mahamba". This lord of the jungle is said to reach a shocking 15m (50 feet) in length!

In the late 19th Century Belgian explorer John Reinhardt Werner reported sightings of giant crocodiles that lend weight to the terrifying folk tales of the native population.

Whilst journeying down the Congo on the Aja, a 12.8m (42 foot) steam launch, Werner stopped at a sand bank to shoot ducks. he shot one and pursued others over a low ridge when he saw: "the biggest crocodile I have ever seen. Comparing him to the Aja which lay in deep water some three hundred yards off, I reckoned him to be quite fifty feet long: whilst the centre of the saw-ridged back must have been some four feet off the ground where his belly rested."

---

* Kearton published three books: *Through Central Africa* (1915), *In the Land of the Lion* (1929) and *Adventures with Animals and Men* (1939). I shall be searching for this picture in future.

Werner stupidly took another shot at the ducks (they had run out of meat on the ship), and alarmed the monster which made off into the water. the creature was also witnessed by a native boy that Werner had with him.

Around three days later Werner saw another vast specimen. The Aja had embedded itself in a sandbank when it was heaved up out of the water by something causing a commotion under the ship...

"I saw an enormous crocodile - longer I am certain than the Aja - rush across the bank and tumble into the deep water beyond. I never before saw such a large crocodile move so fast, and I had no time to get a shot at him. He must have heard us coming and was trying to make for the deep water on our side of the bank, when we ran into him and hammed him onto the sand. We struck him, moving at a rate of four miles per hour, but during the short time he was in view I could not see that he bore any marks of the collision!"

It would be as well now to pause and reflect on the dimensions of such a huge crocodile. A 7.5m (25 foot) creature would be an awesome animal in the two to three ton weight bracket. A fifteen metre (50foot) animal would be a colossal weight. When an animal doubles its size its weight increases eightfold. This is because length, depth, breadth and height have all been doubled. If we take the conservative estimate of two tons for the weight of a 7.5m specimen then an animal fifteen metres in length would weigh in the region of fifteen tons - twice the weight of an average elephant. If crocodiles of these dimensions do exist then they are the largest macro-predators on the planet. Most of the great whales are plankton feeders and even the toothed sperm whale feeds mainly on small fish and squid (the giant squid forms only 1% of its diet and weighs far less than the sperm whale in any case). Such a giant crocodilian would be surpassed only by the giant marine reptiles of the Mesozoic and possibly the largest carnivorous dinosaurs (palaentologist Gregory.S.Paul postulates a maximum weight of twenty tons for the largest Tyrranosaurs). If they do indeed exist there is no animal on earth that could withstand an attack from one of these giant saurians.

It is obvious that we are not dealing with whole races of gargantuan crocodiles, but a few massively large individuals. If they were prehistoric survivors of a giant race then many more survivors would have turned up!

So what is it that causes certain crocodiles to become so large? I believe that it is a combination of several factors.

Both Nile and Indopacific crocodiles have large distributions. Within their range many sub-species can exist and these may display large variations in size. A striking example is found along the Aswa River in Northern Uganda. Crocodiles here reach sexual maturity at between 1.5 and 1.8m (4.9 to 5.9 feet) and never exceed 2.1m (seven feet). This is less than half the average size. This tiny race has an unusually large head 30.5 cm long. This SHOULD yield a total length of 2.13 metres but the Aswa crocodiles fall far short of this. It seems that this is a product of prolonged periods of aestivation; the retarded development being due to inactivity. It would seem that this is a strategy that has been developed to avoid food shortages.

Other areas such as Lake Malawi, The Congo, parts of Tanzania, (such as the Grumati River) and the Semliki River in Uganda/Zaire produce larger than average specimens.

Where populations of these larger than average animals have remained undisturbed occasional freaks will be thrown up within the genetic variation that are much larger than the average. The average man is five foot nine inches tall but a lot of the population exceed this. Most cities have several seven foot individuals and the record human height is eight foot 11.9 inches. A large population of "Big Crocodiles" most of whom would reach five metres (16 feet) could throw up seven or eight metre (23-26 foot) specimens occasionally.

Diet is also a factor. Once it was believed that very big crocodilians were immensely old. It was thought that crocodilians grow roughly twelve inches a year until they achieve the length of three metres (10 feet) when the growth rate radically slows down. By this logic to be immensely huge a crocodile must have achieved a great age.

Actually the greatest authenticated age for a crocodilian is a fifty six year old American Alligator (Alligator missippipiensis) at Dresden Zoo...

EDITOR'S NOTE: A crocodile of unknown species died at Yetkatrinaburg in Russia a few years ago at a reputed age of over seventy but this was never, as far as I am aware, properly authenticated.

However, in a paradox, unlike mammals and birds, reptiles seem to live longer in the wild than in captivity.

A female Nile Crocodile called `Lutemba` lived for many years in a small bay of the Murchison Gulf in Lake Victoria. She was the closest thing to a `tame` Nile Crocodile ever recorded. Natives fed her fish and she came like a dog when her name was called. In the 1920s she became quite a tourist attraction and a source of revenue. She features in Cherry Kearton's film "Tembi". At this time she was about 4.25m (14 feet) long which is an average size for an adult male but quite large for an adult female of this species. She was probably at least twenty years old.

Some reports say that she had lived in the area since the 19th Century and had been used as a Royal Executioner by Ugandan Kings (as, indeed they were used by Idi Amin in the Uganda of the 1970's). This does not, however, seem to fit in with her reportedly placid nature. She disappeared in the mid 1940s at an age of between forty five and fifty., but possibly considerably older if she had indeed lived in the years of the Ugandan kings (pre 1894).

Protein intake seems to have more to do with large size than age. In the early 1970s the Louisiana department of Wildlife and Fisheries made some interesting discovery relating to diet and growth rate in the American Alligator.

Two groups of juveniles were reared on different diets. One was fed coypu flesh and the other fish. Nutritional analyses showed that coypu contained 14.9% crude protein, 2.1% crude fat, 0.1% crude fibre, and 45% moisture.

Fish, on the other hand, contained 9.9% protein, 4.0% fat, 1.0% fibre, and 60.6% moisture. Specimens fed on coypu grew 20% larger than their fish-fed peers over a period of three years. They were also more active and aggressive.

Food with more protein content causes accelerated growth. The Aswa crocodiles were tiny due to aestivation brought on by seasonal food shortages. In other areas where protein-rich food is plentiful the average size of the crocodile population was much greater. So, if we conceive of a population of naturally big crocodiles, feeding on protein-rich prey, that occasionally produces a giant freak whose size is increased still further by its diet, then one can conceive of a truly vast animal.

Open seas and teeming rain forests would offer such an abundance of prey. It can be no coincidence that the largest reported crocodiles are seen in these very habitats.

* * * * *

"Will he make many supplications unto thee? Will he speak soft words unto thee?"

BIBLIOGRAPHY

ALDERTON, D. Crocodiles and Alligators of the World (Blandford 1991).

BLASHFORD-SNELL, J. Mysteries - Encounters with the unexplained. (Bodley Head, 1983)

GUGGISBERG, C.A.W. Crocodiles, Their natural history, folklore and conservation. (David & Charles, 1978).

LEVY, C. Crocodiles and Alligators. (Apple Press, 1991).

PENNY, M. Alligators and Crocodiles (Boxtree, 1991).

ROSS, C.A. Crocodiles and Alligators (Merehurst Press, 1989).

WEBB, G.J., MANOLIS.C., and WHITEHEAD., P. Wildlife Management - Crocodiles and Alligators (Surrey, Bratt and Sons. 1987).

HEUVELMANS, B. In The Wake of The Sea Serpents (1968, Hart Davis) if I ever get the bloody thing back!!!!!

A Poll of Totems

A brief examination of animal totemism in North America

by

Tom Anderson

"Prairie Fire" 1953 by Blackbear Bosin; Kiowa/Comanche.
Philbrook Museum of Art, Tulsa, Oklahoma.

No evidence has ever been found of the existence of the remains of high apes, hominids or neanderthals in the Americas. The most ancient sites unearthed from the limestone are distinctly those of Homo sapiens. Early Australoid or Proto-Causacoid, they hunted the mastodon of the high plains before dying out. This was due either to their prey, or indeed themselves, becoming victims of the glacial ice retreating northwards, or by extermination and absorption by waves of human invaders arriving from Asia. Amerindians can be traced back to the Altai by way of the Siberian-Alaskan bridge before it sank into what is now the Bering Strait. Colonising the vast continents eventually led to a loss of the mongoloid features and by the mid 17th Century there was enormous physical and cultural diversification - the so-called "Five Hundred Nations".

Note: White North Americans coined the term `Native Americans` to describe the indigenous inhabitants. Whether this was borne out of a sense of guilt stretching back two hundred years or out of extremes of political correctness is irrelevant to the parties described. Indians call themselves by the name of their nation - Lakota, Nermuneh, Innuit etc in the same way as would an Italian or a Greek. As a generalisation they use the term "Indian" because they know who they are and are currently undergoing a reinvention of themselves and their ethnology. As each tribe had names appended to them by the Spanish, French and British, as well as their own and those bestowed upon them by men of other tribes I will use, for reasons of brevity, those familiar to most non-students of the subject.

The Amerindians, from the arctic to the Andes, shared a view of the world derived from their Asian ancestry. It did not connect cause and effect, and made magic a surrogate for science.

Prior to the Spanish invasion, two great empires (the Mexica or Aztecs in Central America and the Incas in Peru) had discovered mathematics, devised a calendar superior to the Gregorian, produced engineers and artisans, constructed huge palaces and splendid monuments all without the horse and cow, the wheel or metal tools. All of this was fueled by a religion so strong, that when confronted by the battalions of Cortez, Moctezuma the Younger sent out a band of wizards rather than the thousands of men that he commanded. This faith in higher powers is the common bond that linked a multicultural race composed of everything from crop farmers to nomadic warrior societies.

Common to virtually every nation was the deification of the sun as the supreme being and the giver of life.

The most extreme form of solar worship was that of the Azteccans who believed that the sun would not rise unless propitiated by a smoking heart sacrifice. Nor would the crops grow unless irrigated by human blood. Although a sophisticated structure, the Aztec Empire was a totalitarian regime and inevitably crumbled due to its dependence on a religious elitism and an inability to accept change. This latter was the common failing which led to the near extinction of many Amerindians.

The Indian realised his dependence on the elements and sought to enlist their help by the use of religious ceremonies. Secondary to this was animal worship and a form of anthropomorphism imbuing the totem with human characteristics and assuming theirs in return.

Fig. 1. Athabascan Totem Pole from the Pacific Northwest - Tsimshian.

The eagle was considered most sacred as it represented the Great Spirit. To hold its feather in your hand rendered you incapable of lying and its down was called "The Breath of Life". Prior to the advent of modern conservation laws its feathers were plucked from a living bird caught by hand out of respect for its status. The war bonnets of the plains tribes represented a comet's flight in shape but modern versions are made from died goose or turkey feathers (legal eagle).

The "Flicker" or Red Breasted Woodpecker was considered heroic. Its drumming represented both heartbeats and the beat of the earth. Its blood coloured feathers were offered to water spirits as protection against both human and spiritual enemies and when worn in the hear signified that the wearer was a member of a medicine society. Its birth sign was the 21st June to the 22nd July.

The Innuit creator Tulugaukuk was a raven and equates to Libra in Chipewe astrology.

In common with other cultures the Navajo feared owls as death omens and seeing one required a purification ceremony. The Hopi merely consider the Great Horned Owl (Mongwa) to represent discipline in their complex religious rites. The Hopi are best known for their Chusona snake-handling dances to conjure up rain for their crops. Unique to the south west Zuni tribe are animal fetishes embodying desirable traits of the animal which are used in prayer drawing into the home the power inherent in the animal. The Zuni are famous for their silversmithing, often using turquoise in their jewelry. I wear a sunray ring which supposedly brings luck and is a strong love charm. I have reason to believe the first part, though the second has proved somewhat less than effective to date.

Designed by Black Elk, the Lakota Ghost Dance shirt was made of muslin, dyed to resemble buckskin and considered bulletproof. This example bears a Thunderbird motif. The Ghost Dance was conceived by Wovoke, a Piaute and its implications led to the assassination of the Hunkpapa shaman Sitting Bull (Tatanke Yatonka) and three hundred people at Wounded Knee in 1890.

Fig.2. Sioux Ghost Dance shirt

The best known totem is the Thunderbird, interpreted in various ways usually with lightning flashing from its beak, and with its wing beats representing thunder. Its work is to water the earth and is often pictured alongside eagles and falcons.

Geographical location dictated a nation's totems. Coastal northwestern tribes respected the mountain goat for its nobility, the grizzly for strength and the Orca as ruler of the seas and the Underworld. This area is known for its totem poles, usually carved from cedar, unique to each family, and so treasured that they took them inside in bad weather.

Many tribes call North America "Turtle Island", their traditional belief being that the earth is borne on the back of a turtle.

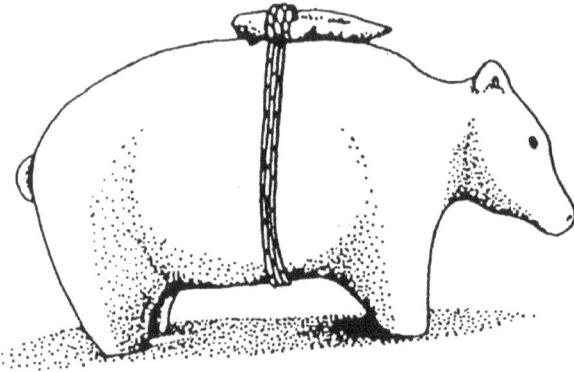

Fig. 3. Stone Carved Zuni Bear Fetish.

Fig. 4. Sun Bear, Chippewa shaman, author, lecturer, and collator of totemic star signs in text.

The bear was thought to be head of the animal council for his strength and courage and in most tribes the Bear Clan was the medicine or leadership group. Bear claws were commonly worn for power, the animal simultaneously venerated and killed as is the case with the Japanese Ainu people.

Mad Bear Andersen was an Iroquois shaman, and as far as I know no relation.

Because of the plain's tribes dependence on it, the buffalo featured highly. The Pawnee legend has it that the Milky Way is the dust cloud following the race between a buffalo and a horse. The Navajo call it "The Rainbow Bridge" which connects the earth and the sky.

Some animal totems are less obvious.

The Zuni revered the bear, cougar, wolf and badger for their courage, the snake for its patience and the shrew, oddly enough for its energy. Unfortunately for the editor of this book who spent many weeks during 1996 pursuing Amerindian legends of Singing Mice, (for a paper published in Steve Moore's Fortean Studies Volume Three) I have been unable to discover any.

Hopi tadpole emblems on bullroarers signified life power (e.g. male sperm).

Those observing the correct rites travelled "The Good Red Road", signifying the right way to live. Those who broke the taboos risked encountering the Underworld. Crop Failure was blamed on the Cloud Swallower who caused drought and famine. Iktomi the spider man visited evil on the unwary, but worst of all was being kidnapped by the Nakani, supernatural humanoids of the northern forests who were reputedly capable of driving a man insane.

Another good way of risking madness was to overdose on Mecal (Sophora secundiflora), derived from the peyote cactus button and used for hallucinatory visions. Four of these 'buttons' was considered a safe limit, the rest to be worn on a necklace until required.

Fig.5. Medicine Wheel.   Symbolic of the "Circle of Life" it also had calendric and symbolic significance.

Quanah Parker, the last great Comanche Chief, championed the Peyote Religion which is now incorporated into the First American Church.

Fig. 6. Quetzalcoatl Temple Relief c.10 A.D.

Quetzalcoatl reputedly lived circa 940-1070 A.D. and is depicted in petroglyphs as a man carrying quetzal feathers. Translated from early Mexica as "Plumed Serpent", later examples are actual snake carvings as he took many forms and is a Mayan messiah prophesied to return in various guises.

The most famous ceremonial was the plains tribes sun dance, (Waiwaugag Wachipi) of the Sioux, held during the full moon. A lodge was built around twenty-eight poles to represent both the lunar cycle and the number of ribs of the buffalo. The supplicant's back muscles are pierced by skewers tied to buffalo skins which are dragged around until the skin tears. This can bring on fainting and resultant visions which are then interpreted. Adopted by Cheyenne and Ponca it was condemned by the Christian Church but re-emerged during the second world war and is still practiced as a rite of passage in defiance of Federal law.

A less traumatic ceremony to taunt the white man is the Cherokee "Booger Dance". This involves wearing caricatured anti-causacoid masks with names like "Pale Faces who pee in the woods".

Fig. 7. Ceremonial
carved wolf mask-makah,
Washington State.

The Nations have been subjected to genocide, degradation and Disneyfication to intolerable levels but have retained an affinity with the natural world that we can only envy.

For us, rather than them, it is perhaps too late!

The Fortean Fauna of Percy Fawcett

Cryptozoology and zoomythology in the Records of Lt-Col. P. H. Fawcett

by

Mike Grayson

The name of the explorer Percy H. Fawcett will probably be familiar to readers of A&M in one particular context: his reported encounter in 1907 with a giant anaconda, whilst on the Rio Abuna at Bolivia's northernmost frontier with Brazil.

This episode has become a 'classic' of cryptozoology, and the details have been repeated in more than one book; probably the best known being Heuvelmans' (1). However, amongst the letters, log-books and other records of the explorer - later arranged and published in book form by his son, Brian Fawcett(2) - are many other notes on natural history. These are often of interest to the cryptozoologist and Fortean zoologist.

First though, it may be appropriate to say a few words about the man himself. A career in the army that included learning the art of surveying whilst in Malta led, in 1906, to the offer of carrying out boundary delimitation work in Bolivia. The frontiers between Bolivia, Brazil and Peru were at that time ill-defined, and the Bolivian government had requested the Royal Geographical Society to supply an impartial 'referee'.

From this beginning, Fawcett made several expeditions in South America between 1906 and 1925; firstly as a boundary surveyor, but later as an explorer fuelled chiefly by his belief that in southern Brazil were hidden the remains of an ancient and advanced civilisation. This quest was to lead eventually to his disappearance in the Matto Grosso. Along with his eldest son Jack, and the latter's friend Raleigh Rimell, Fawcett vanished into the wilderness: his last communication was dated 29 May 1925.

Whatever else he may have been, Fawcett was not a trained biologist, nor was South America's rich variety of fauna and flora any part of his reason for being on the continent. His own knowledge of animals appears not to have been great, as he makes the occasional 'howler' - for example, he refers to llamas as "those proud and dignified relatives of the sheep"(!). Probably most school-children today would know that llamas are related to camels rather than sheep, but I mention this not to have a cheap jibe at Fawcett: merely to indicate that Fawcett's notes on the fauna he encountered are not backed up by any personal expertise in this field.

He sometimes repeats information that must have come to him at second- or third-hand, and accepts it in good faith, whereas we may in some cases be looking at local folklore rather than hard zoology. Bearing these provisos in mind, we will now go on to look at the various categories of cryptic wildlife mentioned in the records of Percy Fawcett.

1) <u>The Big Black Snorer</u>   Fawcett's fatal encounter with a giant anaconda (fatal for the anaconda, that is) is too well known to warrant a detailed retelling. Suffice it to say that, having shot the reptile as it slithered up a riverbank, Fawcett estimated its length as 62 ft.(19 metres): fully twice the maximum length accepted by most authorities for this species.

Though this measurement was an estimate, it was made at very close range and it is hard to accept that any major error was made. Nor do I think that Fawcett was a bare-faced liar. He certainly had nothing to gain by making a false report of this kind.

Following his account of this episode, Fawcett goes on to give other information on over-sized anacondas. He states "In the Araguaya and Tocantins basins there is a black variety known as the **Dormidera** or 'Sleeper', from the loud snoring noise it makes. It is reputed to reach a huge size, but I never saw one."

The two river basins mentioned here are in eastern Brazil, far to the east of Fawcett's own encounter on the Abuna. These giant snakes must therefore have quite an extensive range.

But what of the supposed 'snoring' sound ? In some ways, this feature is more surprising than the size of the snakes. Beyond hissing sounds, these reptiles are usually regarded as almost mute. Many anacondas have been kept in captivity, but has anyone heard such noises coming from zoo specimens ? It is worth mentioning here that there is another crypto-snake, on another continent, which is also reported to show vocal talents quite beyond the usual repertoire: the crested crowing cobra of Africa(3).
As we shall see below, Fawcett gives a personal report of another species of snake making unexpected noises, so we shall look at this subject again. Meanwhile it is worth saying something about the Dormidera's coloration.

Normally, anacondas have an olive or brownish-green background colour, patterned with black, oval-shaped markings. If the 'Dormidera' is all black, as reported, might it be a totally different species rather than an out-sized anaconda ? Interestingly, more than 20 years after Fawcett's disappearance, an incident took place in the Araguaya basin which might shed some light on this question.

In 1947, members of an expedition seeking to establish peaceful relations with the Chavantes tribe came upon a huge snake. One expedition member later gave an account to Heuvelmans(1), telling how the snake was killed and measured. It was even longer than Fawcett's specimen, being between 22 and 23 metres long. It was not black, but seemed to be an unusually dark specimen of anaconda: its background colour was said to be "very deep dark brown." Here then seems to be some confirmation of Fawcett's information that in the Araguaya area, anacondas do reach a huge size and (sometimes at least) a dark coloration.

In more recent decades, reports of such enormous serpents seem to have become fewer. One wonders whether the Dormidera still snoozes peacefully by some foetid Brazilian swamp, or has it now entered the Long Sleep of extinction ?

2) Rattlesnakes Galore ?   Venomous snakes were a constant hazard to
Fawcett and his companions as they made their way through the wilderness. It is perhaps then not surprising that snakes feature more often
than any other type of fauna in his writings. Several forms are mentioned which may represent new species, but these claims need to be carefully
assessed as things are not always as straightforward as may first appear.

Speaking of the Beni River region of Bolivia, Fawcett notes that it is
the haunt of various kinds of poisonous snakes. "Commonest is the rattlesnake. There are five different kinds, but in length they seldom exceed
a yard." This information does not agree with current herpetological
knowledge, as only one species of rattlesnake would be expected to occur
in Bolivia. This is the species known scientifically as **Crotalus
durissus**; a form which has an average adult length of about 4 feet.

Do we then assume that there must be several small species of
rattlesnake awaiting discovery in northern Bolivia? Perhaps this
is indeed the case, but it would be unwise to accept this without
looking at other possibilities.

Firstly, I think it is rather unlikely that Fawcett ascertained
personally that five distinct forms of rattlesnake were present in
the Beni area. Probably he is here repeating local opinion. We
should remember that local names for animals do not always match
with scientific views on taxonomy. Distinct names may be given to
different sizes or colour variations of the same species; or even to
the same species encountered under different circumstances. (An
example of the latter is the name 'coatimundi', applied to solitary
individuals of that normally-gregarious relative of the raccoon, the
coati.)
Another possibility is that the term 'rattlesnake' itself is being
used very loosely. South America has many species of snake of the
Fer-de-lance group [**Bothrops** & related genera]: these belong to the
same zoological family as the rattlesnakes, but lack the distinctive
rattle on the end of their tails. However, some of these snakes are
known to vibrate their tails rapidly when they feel threatened, and
if this occurs amongst foliage or forest-floor litter it can produce
a warning sound.

When we take the above into account, it's clear that mention of "five
different kinds" of rattlesnake does not necessarily constitute proof
of new species awaiting the intrepid explorer-naturalist.
The Truth, as we all know, is out there: but cryptozoologists must be
prepared for it sometimes to be a rather prosaic truth.

3) VERY Strange Snakes   If the previous section was a little tame for
aficionados of the weird, then maybe the following will be more to
their taste. From the Rio Abuna region, Fawcett reports "One species
of snake here had a head and one third of its body flat as a tape,
while the rest of it was round." Stranger still- "A number of people
assured us most emphatically that in this region was a snake about
three feet long which telescoped into itself before striking. As

though to modify the account of what sounded to me like an anatomical impossibility, they added that it was not very poisonous. I should like to have seen one, but never did."

The sceptical might say that there was an obvious reason why Fawcett didn't see the 'Telescope Snake': that it had no existence outside of local legend. It is true that few creatures have generated as many myths, lies, superstitions and exaggerations as have snakes. Fawcett himself records the odd South American belief "that a little bag of bichloride of mercury carried on the person will prevent snakes from attacking." Perhaps the Telescope Snake's link to reality is as tenuous as beliefs such as this.

I would, however, suggest a tentative basis for reports of this odd-sounding cryptid: could this 'snake' actually be a giant earthworm? This may at first appear an unlikely concept, but the reported size of this snake is no obstacle to a 'worm' explanation. The earthworm **Glossoscolex giganteus** of Brazil, for example, can reach four feet in length when naturally extended; and this is not South America's only known species of giant annelid.

Could the great elasticity of these worms, with their ability to stretch (or be stretched) and contract, provide an explanation of a snake-like creature that "telescoped into itself"? Of course, these earthworms are entirely harmless and are hardly likely to 'strike' at anything: on the other hand, any yard-long legless animal might attract a certain amount of local mythology. If any A&M reader has other explanations for the Telescope Snake, I would be interested in hearing them.

There is one further strange serpent to consider: one whose local name means the 'Fire Extinguisher'. This seems to be a form of Bushmaster, which is the largest known venomous snake of the New World, reaching a length of upto 12 feet. In his writings, Fawcett often refers to it by its South American name of SURUCUCU. As well as the normal form, he also makes mention of the **Surucucu Apaga Fogo** (Bushmaster that extinguishes fire). These are said to be "attracted by fires, and forest men are so much in fear of them that they never keep their fires alight at night." They are supposed to "coil up on the ashes of a smouldering fire", but have yet another attribute unusual in a snake: "They are said to have an extremely keen sense of hearing" reports Fawcett.

Now as all known snakes lack an external ear, an eardrum, a tympanic cavity, etc. the sense of hearing is not one of snake-kind's strong points. Traditionally they are viewed as being deaf to air-borne sound, though they can pick up vibrations in the ground. Interestingly, since Fawcett's time, some experiments have indicated that snakes can detect sound - especially within a frequency range of c.150 - 400 Hz. [see brief mention of these experiments in Reference (3)]. However, one must wonder why the 'Fire Extinguisher' of all snakes, was credited with a "keen sense of hearing".

The question remains as to whether the Fire Extinguisher is a different species, rather than a Bushmaster given mythologized attributes. It is certainly true that the Bushmaster has attracted various legends: in some parts of South America it is said to suckle from cows, and from sleeping women. The snake as milk-stealer is a widespread myth, told in various parts of the world about various species of serpent, but without any basis in truth. (In fact, snakes in captivity totally refuse milk when it's offered to them). The habits of the Fire Extinguisher could likewise be only the stuff of legend, but I think we can't rule out the possibility that a second form of Bushmaster exists, which the local inhabitants distinguish from the known variety.

4) Do Bushmasters Whine, & Anacondas Wail ?   In the above section, we were looking at information reported to Fawcett, but not personally verified by him. In this section, though, we look at one of Fawcett's own experiences, which takes us back to the subject raised briefly when we considered the Dormidera: the topic of snakes making unusual and unexpected vocalisations.

Let us quote Fawcett's own words, as he and a local set off into woodland on a hunt for game. ".. as we passed a tree with a hole in it about ten feet from the ground we heard a thin, shrill whining noise. My host clambered into the tree well above the hole, and emptied his shotgun into the cavity. Like a jack-in-the-box a surucucu shot out into the air, fell to the ground, and scuttled away into the undergrowth. Had either of us been within reach we would surely have been bitten; as it was, the shock was unpleasant. I now had evidence of the truth of the story that these reptiles whine when sleeping - it reminded me of anacondas which give voice to melancholy wails by night, a weird sound I have heard scores of times."

It is possible to argue that the whining sound was not made by the Bush-master, but by some other inhabitant of the tree-hole: perhaps some small mammal that was the snake's intended prey? Most textbooks do not credit snakes with much in the way of a vocal range, but there is some evidence that they are capable of more than is generally accepted. Karl Shuker, when discussing the African crowing crypto-cobra (3)gives various reports of known species of snakes apparently emitting purrs, coughs, bleats and other noises. One of the most startling, but at the same time most convin-cing, reports came from members of an expedition to Sarawak in 1980. In the darkness of a cave, they heard an eerie "yowling miaow". Their torch-light revealed the vocalist to be a non-venomous species of snake, called the cave racer. Until that time, no one had even suspected that the racer was capable of more than a hiss. The purpose of such a strange call is still unknown.

In the light of accounts such as this, we should not be too quick to dismiss the concept of Bushmasters making a whine. It is interesting that local people claimed this sound was made when the snakes were sleeping: it would not be to an animal's advantage to draw attention to itself at such a vulnerable time. Certainly this episode suggests a fascinating possible avenue of research for some open-minded herpetologist.

As for wailing anacondas... were the sounds Fawcett heard "scores of times" really made by these constrictors ? Perhaps he accepted local folklore in attributing strange nocturnal cries to anacondas ? He does not claim to have actually caught an anaconda in the act of uttering these melancholy calls. But after the case of the Cave Racer, who can say for sure ?

On this note, we will leave the subject of snakes and move on to other areas of the Animal Kingdom, beginning with:

at first sight to be the form referred to by Fawcett when he writes of the Abuna River: "Here, too, you see the bufeo, a mammal of the manatee species, rather human in appearance, with prominent breasts." After reporting that it is regarded locally as good eating, he makes the surprising statement "It is neither helpless nor inoffensive, and will attack and kill a crocodile."

Now 'inoffensive' is exactly what most people do consider manatees to be, and the thought of one attacking a crocodile is hard to envisage. Yet later in his records, Fawcett claims to have witnessed just such an event: "Vast quantities of camelote, or sudds, held us up on the river between here and Porvenir, on the Paraguá. On the way we witnessed a fight between a manatee and a crocodile. One doesn't expect much fighting ability in the humble sea-cow, but it whipped the crocodile decisively."

There is one other reference to the bufeo in Fawcett's writings. When at the mouth of the Rio Verde, "Where there was free water bufeos gambolled round the boat". It is no more easy to visualize a manatee 'gambolling' than to imagine one savaging a croc. Surely this behaviour sounds more like that of dolphins ? And here, indeed, seems to be the solution to this odd matter: Fawcett appears to have confused two very different types of aquatic mammal – the manatee and the river dolphin.

Known scientifically as **Inia geoffrensis**, the South American river dolphin is known in Brazil by the name 'boto' or 'bouto'. The Spanish equivalent of this Portuguese word is **bufeo** (6); the term used by Fawcett. This cetacean is known to bow- & wake-ride with boats, and there can be little doubt that this was the species seen by Fawcett.

An apparent mystery of animal behaviour – the placid vegetarian manatee attacking crocodiles – is thereby resolved, and shown to be based on a misidentification. Even so, Fawcett's observations are not without zoological interest: if river dolphins do sometimes attack and kill crocodilians [presumably caymans, rather than true crocodiles], rather than simply avoiding them, this piece of behaviour would merit further investigation.

7) Ihe Return of Draculae ? An almost casual comment of Fawcett's regarding vampire bats just might have greater significance than is immediately apparent. Whilst in the forests of South-eastern Peru in 1910, Fawcett's party suffered from the unwelcome attentions of these bats. He reports how he felt one of the vampires at work on him one night, and says "The large and small varieties use the same tactics." But exactly which 'varieties' is Fawcett referring to ?

There are, in fact, three known species of vampire bat; all from Central and South America. The common species (Desmodus rotundus) is the type associated with attacks on humans and livestock, and must be one of the varieties that beset Fawcett.

Of the other known species, the White-winged Vampire (Diaemus youngi) is said to only attack birds; whilst the Hairy-legged Vampire (Diphylla ecaudata) also appears to prey mainly on birds, but may attack some mammals too. It is smaller than the Common Vampire, but all 3 species are only 2½-3½ inches long. Does any of them count as a 'large' variety ?

One must obviously be cautious here, for fear of creating a mystery where none truly exists. Perhaps Desmodus is Fawcett's larger variety, and **Diphylla** the small one. Perhaps the latter is more prone to attack mammals than is currently believed, or maybe its habits vary from location to location. Perhaps once again Fawcett repeats some scrap of folklore concerning the existence of two types of vampire, and didn't see himself that two clearly different forms were involved. Perhaps...

In his   recent book, Karl Shuker (7) mentions reports of large vampire bats from the Ribeira Valley of southern Brazil, which may correspond to the (supposedly extinct) Pleistocene species **Desmodus draculae**. We would expect this larger relative of the Common Vampire to seek its bloody meals from mammals, rather than birds. So is it just possible that Fawcett's short comment is a pointer to the survival of **draculae** into the present century ? An intriguing thought, but so brief and undetailed is Fawcett's note that we can do nothing but speculate.

8) The Stone-pecker and The 'Acid Leaf'   South America is incredibly rich in bird species, but Fawcett makes relatively few mentions of birds in his records, and it seems that ornithology did not hold much interest for him.
One curious account that involves a bird is really more a question of cryptobotany than cryptozoology. Fawcett refers to "a small bird like a kingfisher" in the forests of Peru and Bolivia, which makes its nest in round holes in rocky escarpments above rivers. An informant told Fawcett that the birds made these holes in the hard surface by rubbing a kind of leaf onto the stone with their bills. After several applic- ations, some quality in the leaves softened the stone enough for the birds to peck out pieces of it.

Fawcett himself regarded this story as a 'popular tradition', after hearing similar reports from other sources.
It is perhaps not useful to dwell on the possible identity of the bird species involved, given that we are once again faced with so few details. There are members of the Kingfisher family in South America, as well as less familiar forms which bear a superficial resemblance, and that also nest in burrows in banks, such as the Jacamars. Any of these – or maybe more than one species – could be the bird(s) involved. The real mystery here, though, is the plant used by the birds. Its reported properties are certainly remarkable, but Fawcett may have been correct in believing that this wonder-leaf grew only in the human imagination.

9) The Parliament of Condors   For those interested in birds, Fawcett's stories about condors offer food for thought. In general, though, he seems to repeat information given to him by one Carlos Franck; a German-Bolivian with whom Fawcett stayed whilst travelling in the Andes. Señor Franck apparently "knew these mountains like the palm of his hand", but some of the accounts he passed on to Fawcett would certainly raise the eyebrows of modern ornithologists.

"They rarely come down below fifteen or sixteen thousand feet, except to carry off a sheep or - and there are authenticated cases of it happening - a child." One would like to know more about these 'authenticated cases', but even more amazing news follows: "near Pelechuco a full-grown man was carried about twenty yards" by a condor!

Stories such as these totally contradict our accepted knowledge of condors. These birds belong to the group known as Cathartids, which although outwardly resembling the vultures of Africa & Eurasia, are actually closer to the storks in phylogenetic terms. One result of this relationship is that the talons of cathartids are weaker than those of vultures and eagles. Zoologists would not expect these birds to be capable of carrying off sheep... or humans! But even the above tales pale in comparison with Carlos Franck's account of a Condor Council. It is an interesting addition to the Fortean literature on animal courts, parliaments and battles: I will quote Fawcett's version in full-

"[Carlos Franck] once came upon a council of king condors. A large ring of solemn birds surrounded two huge black ones and one still larger white one which seemed to be the leader. He had long wanted one of the rare white condors as a trophy and was unwise enough to shoot at it. At once the circle of birds broke up, and two immediately set on him, so that he was forced to throw himself on his back and beat at them with the rifle as they swooped. He escaped, but they followed as he scrambled down the narrow, rocky path on the sheer mountain-side, trying whenever possible to knock him off with their wings into the abyss. He considered himself lucky to have escaped."

Although Franck claimed to be the protagonist of this incident, and Fawcett seems to have believed it, to me it has all the flavour of a piece of folklore. Not only are his condors engaged in unusual behaviour, holding a 'council', but they have a leader of unnatural size and colour. It may be worth pointing out that Franck also told Fawcett another incredible story: of how Franck's daughter, who was crippled by hip disease, and whom four operations had failed to help, was totally cured within a week after taking a potion prepared by a local Calahuaya [native herbalist].

Perhaps I do Carlos Franck a severe injustice, but on reading all this, I must admit to wondering whether he wasn't just the chief bullshit-merchant of the Andes. However, out of fairness, I will let Señor Franck have the last word (as recounted by Fawcett):

"Living in these isolated places, very close to Nature and away from the rush and bustle of the outer world, one experiences many things which an outsider might consider fantastic, but which to us are commonplace."

10) <u>GUMS! Toothless Terror of the Paraguay</u>   Surely there could be few fates more awful, and yet at the same time vaguely ignominious, than being gummed to death by a toothless shark. And yet this dire destiny may have befallen hapless swimmers in the Paraguay River. For here, says Fawcett "there is a freshwater shark, huge but tooth-less, said to attack men and swallow them if it gets a chance."

Now there is one species of shark, known to attack humans, which habitually enters freshwater in the warmer regions of the world. This is the Bull Shark (Carcharhinus leucas), which has been recorded far up the Amazon, as well as in rivers such as the Zambezi and the Mississippi. Growing normally to around ten feet in length, the Bull Shark is large - though hardly 'huge' - but is very far from being toothless. For this reason, I don't feel we can dismiss Fawcett's mystery fish as having been explained away by wandering Bull Sharks.

The general description of this gummy terror fits another group of fishes rather well: the sturgeons. These are toothless, have a shark-like outline, and some species are known to reach a great size. Even the largest sturgeons, however, would not "attack men and swallow them". We could perhaps put the man-eating habits down to folklore, but the idea of a sturgeon identity for the Paraguay 'shark' falters at another hurdle: all the known species of sturgeon occur in the Northern Hemisphere, in North America and Eurasia. A South American sturgeon is as unlikely as a toothless Bull Shark.

Where this leaves us - other than puzzled - I'm not quite sure. Despite having a rich variety of aquatic exotica, including freshwater dolphins and stingrays, South America has no known inhabitant that can fit the description of the 'toothless shark'. Either the region's biggest freshwater fish is still awaiting scientific discovery, or else folklore - possibly aided by the occasional wandering bull shark - has created a phantom fishy terror.

11) Here Be Dragons...

We have looked at strange snakes, stone-pecking birds and a fearful fish, but what of **real** monsters? What of possible primeval survivors from an earlier epoch, lurking in some last lost corner that Time obligingly forgot? Fawcett was in many ways the archetypal explorer, relentlessly pushing into areas where, even in the 1920s, the maps were largely blank. Those areas where the early cartographers had written "Here be Dragons" to mask their ignorance. Did Fawcett find any dragons in his travels? The answer seems to be... yes and no.

Certainly Fawcett never claimed to have seen any monsters personally, but he had heard stories which he clearly gave some credence to. He speaks cautiously on such matters; perhaps wary of further ridicule from those who dubbed him a liar for his encounter with a giant anaconda. But still, he speaks:

"They talk here of another river monster - fish or beaver - which can in a single night tear out a huge section of river bank. The Indians report the tracks of some gigantic animal in the swamps bordering the river, but allege that it has never been seen." This refers to the Paraguay River, and the quote appears in the same paragraph as the mention of the toothless shark that we looked at earlier. He goes on to state that "there are queer things yet to be disclosed in this continent of mystery" and concludes with the following sentence:

"In the Madidi, in Bolivia, enormous tracks have been found, and Indians there talk of a huge creature descried at times half submerged in the swamps."

Fawcett returns to the subject of the Madidi Monster later in his journal, though few extra details come to light:

"..and in the forests of the Madidi some mysterious and enormous beast has frequently been disturbed in the swamps - possibly a primeval monster like those reported in other parts of the continent. Certainly tracks have been found belonging to no known animal - huge tracks, far greater than could have been made by any species we know."

Not for the first time, one finds the lack of particulars in Fawcett's writings very frustrating. He gives no description of these 'huge tracks', nor does he say where he heard about these supposed monsters. Did he speak with any eye-witnesses? Did he obtain accounts of what these beasts actually look like? Alas, we will now never know how much - or how little - Fawcett really learned about these South American 'dragons'.

In a way, these hints and half-accounts of monsters are typical of the stories that have emerged from South America concerning 'living dinosaurs'. The accounts are neither as numerous nor consistent as the reports from Central Africa regarding 'Mokele mbembe' and its cousins. Perhaps the Madidi marshlands still hide some cryptozoological surprises, but the nature of such maybe-monsters remains open to a great deal of doubt.

*

This concludes our survey of Percy Fawcett's writings insofar as they touch on our favourite subject. Those who are interested also in such topics as lost civilisations, or the study of 'primitive' cultures, will find much food for thought in these writings.
For myself, I rather hope that some of the species of odd wildlife he came across will one day find their way onto zoology's ACCEPTED file: that we will watch the mitla on TV Natural History programmes, or view the dormidera in the zoo's Reptile House. If so, perhaps some scientist will have the grace to apply the name 'fawcetti' when describing such a new taxon, in honour of a remarkable and unique individual.

*****

REFERENCES

(1) HEUVELMANS, Bernard: On the Track of Unknown Animals. Rupert Hart-Davis: London, 1958.
(2) FAWCETT, Lt.-Col. P.H.: Exploration Fawcett. Hutchinson: London, 1953.
(3) SHUKER, Karl P.N.: Extraordinary Animals Worldwide. Robert Hale: London, 1991.
(4) MACKAL, R.P.: Searching for Hidden Animals. Doubleday: Garden City, 1980.
(5) SHUKER, Karl P.N.: Mystery Cats of the World. Robert Hale: London, 1989.
(6) WATSON, Lyall: Whales of the World. Hutchinson: London, 1981.
(7) SHUKER, Karl P.N.: In Search of Prehistoric Survivors. Blandford: London, 1995.

Sea Serpents, Scientists and Sceptics

An introduction - by Richard Freeman

There can be few greater accolades for a zoologist than to discover a new species. The fact is that, even in this overpopulated, polluted world of the late 20th century; mankind has still not totally subdued nature and startling new discoveries are still to be made. Here, indeed, be dragons. Cryptozoology, in a sane world, would be the vanguard of the zoological canon.

Sadly, the reverse is true. Academia seems full of arrogant, blinkered "experts" who, despite never leaving their cosy labs and lecture halls, proclaim from their armchairs that this or that cannot exist or we would have known about it. I have no time for armchair zoologists. These appear, however, to be an ubiquitous breed. Grover Krantz once mentioned that if he had a bigfoot corpse, he would have to drag it around to each "expert" and rub his face in it! Like rats, these blind men have been around for a long time. Take the following article on sea serpents, written over 100 years ago. This piece quotes encounters with specimens that rear up out of the sea and spout water; others which react aggressively when shot at. Yet despite this, contemporary zoologists took the easy way out, explaining the animal manifestations as giant strands of floating weed. Some weed that can perform feats like these!!

A couple of cases did turn out to be weed; but a couple of swallows do not make a summer. The kind of thinking behind such an explanation is sloppy and lazy. Cryptozoologists aren't asking science to glibly believe in every strange tale; but if the eyewitnesses build up in their thousands, as with sea serpents, serious investigation is called for. Everyone makes mistakes, but for scientists to think of all laymen as idiots is a grave error.

Unfortunately, we do not know where or exactly when the following article was published, or who the author was. If anyone can provide this information, then please let us know!

CHAPTER X.

THE SEA-SERPENT.

" A monster,
Bred of the slime, like the worms which are bred from the muds of the Nile-bank,
Shapeless, a terror to see; and by night it swam out to the seaward,
Daily returning to feed with the dawn, and devoured of the fairest,
Cattle, and children, and maids, till the terrified people fled inland."

REV. C. KINGSLEY.

WHILE we are discussing the fantastically horrible inhabitants of the ocean-depths, it seems to me desirable that I should devote a chapter to the most celebrated among them, the famous Sea-Serpent, which is at least the cousin-german of the kraken, and generally confounded with the latter in the maritime traditions of the North. A French naturalist—M. Lecouturier—has contributed to the *Musée des Sciences* an excellent monograph on this subject, from which most of the following details have been borrowed.

The fabulous history of the Great Sea-Serpent ascends, like that of the giant polypi, to a sufficiently remote antiquity. Pliny and Valerius Maximus both describe an amphibious serpent swimming in the shallow shore-waters, and only sailing out to sea when he had grown to such dimensions that movement became impossible for him, or, at all events, very difficult, anywhere else than in mid-ocean.

A French author, Belleforest, in his "Cosmographie," comments on the passage in Pliny referring to this marine serpent, and does not hesitate to furnish the most circumstantial details respecting it. According to him, though of colossal dimensions, it was gifted with

* *Musée des Sciences*, 2nd année, tome ii., 1857–8.

ANCIENT FICTIONS.

extraordinary agility. It flung itself on barks and small ships, cap-
sized and dashed them in fragments by striking them with its huge
tail, and afterwards swallowed all their crews. Belleforest adds, with
admirable simplicity, that if the ship was too large for the creature
to crush it, it drew, or rather propelled it towards the shore, in
whatever direction the wind blew ; then waited patiently until the
seamen, compelled by hunger or in the hope of escape, ventured upon
deck or attempted to gain the shore. That was the moment for it
to pounce upon them and crush them with its teeth—for teeth it had,
according to Belleforest. It had also the head of the wolf-dog, with
ears pricked back behind. Add to this a body covered with yellowish
scales, and a croup curving in tortuous folds, and you will have an
exact portrait of the monster; the same, in all probability, which
Neptune stimulated to devour the son of Theseus.

In the north of Europe, a belief in marine creatures of strange
form and prodigious dimensions is widely spread and deeply rooted
in the mind of the masses. As for investigating the exact dimensions
and species of the animals, I need only explain that fishermen and
sailors are too careful of their safety, and that as soon as they descry
one, they think only of effecting their escape with oars or canvas.
Hence the confusion which they make between the kraken properly
so called, or gigantic polypus, and the great sea-serpent, designating
both by the name of kraken, and liberally attributing to them the
most *bizarre* and incompatible characteristics and forms.

"Norway," says Lecouturier, " has an unconquerable faith in the
reality of the great sea-serpent, and ascribes to it the Northern Seas
for a dwelling-place. Pontoppidan, Bishop of Bergen, says that the
Norwegians cherish so strong a belief in the actual existence of this
monstrous reptile, that whenever he spoke of it in a dubious manner,
his listeners broke into a quiet laugh, as if he had doubted the existence of the eel or any other common fish. The name of the ocean-
serpent in these regions is the *kraken;* they also refer to it under the
name of *soe-trolden* (' scourge of the sea ')."

"The Norwegian fishermen," says Pontoppidan, "all affirm,

without the least contradiction in their accounts, that when they sail several miles out to sea, particularly during the hottest days of the year, the sea seems suddenly to grow shallower under their boats; and if they drop the lead, instead of finding eighty or a hundred fathoms depth, it often happens that they obtain scarcely thirty. It is a sea-serpent which thus interposes between the depths and the surface waves. Accustomed to this phenomenon, the fishermen cast their nets, assured that there will be in those parts an abundance of fish, especially of cod and ling, and draw them in richly loaded. But if the depth of the water continues to decrease, and if this movable and accidental shallow rises higher, the fishermen have no time to lose; it is the serpent awakening and moving, rising up to respire the air, and extend its huge folds in the sun. The fishers ply their oars lustily, and when at a safe distance they see, in fact, the monster, which covers a mile and a half of ocean with the upper portion of its back. The fish, surprised by its ascent, flutter a moment in the humid hollows formed by the protuberances of its external envelope; then from the floating mass issue numerous spikes or shining horns, which rear themselves erect like masts crossed by their yards. These are the arms of the kraken."

[Here, then, is a resuscitation of the kraken, the serpent trans-forms itself into a polypus: it has arms, and what arms! Such is their vigour, that if they seize upon the rigging of a ship of the line, they will infallibly capsize her!]

" After remaining some time on the waves, the monster redescends with the same slowness, and the danger is not less for the vessel which may be within its range; for, while sinking, it displaces such a volume of water as to occasion whirlwinds and currents not less terrible than those of the famous Mäelstrom."

" Such is in Norway," continues Lecouturier, " the popular belief respecting the sea-serpent. The old Scandinavian writers, on their part, attribute to it a length of 600 feet, with a head closely resem-bling that of the horse, black eyes, and a kind of white mane. According to them, it is only met with in the ocean, where it suddenly rears it itself up like a mast of a ship of the line, and gives vent to hissing noises, which appal the hearer, like the tempest roar. The Norwegian poets compare its progress to the flight of a swift arrow. When the fishermen descry it, they row in the direction of the sun, the monster being unable to see them when its head is turned towards that planet. They say that it revolves sometimes in a circle around the doomed vessel, whose crew thus find themselves assailed on every side."

We read in Hans Egidius' narrative of his second voyage to Greenland, that, in the month of July, an animal reared its head above the waves to about half the height of the mainmast. This head terminated in a long pointed muzzle, and—what hitherto had not been related of any sea-serpent—it ejected the water through a single vent placed on the summit. In the guise of fins the monster had immense ears comparable to those of an elephant, which it agitated like wings, to keep the upper portion of its body above the waves. It dived after a while by flinging itself backwards, and made a kind of somersault, which showed in succession all the other parts of its bulk covered with large scales.

In this new species of sea-serpent, with its vent and its fin-like wings, we think it possible to recognize another fantastic animal, the "great white whale" of the Greenland coasts, hunted for two centuries by the Scotch whalers, which they called *Maby Dick*, and regarded as the terror of the Arctic Seas. According to these mariners, it makes its appearance now at intervals; but is of so venerable an age that its body is completely covered by vegetation, algæ, and marine mosses, in whose midst live attached to it, as to a rock, multitudes of shell-fish and polypi.

The traditions of the North speak also of a marine monster which was stranded one day on the beach of one of the Orkney Islands. It is said to have measured eighty feet in length and fourteen feet in circumference, to have worn a long bristling mane, which, luminous in night and shadow, grew dull and dark during the day. Despite the fantastic character of some of those details, we may add

that their general veracity is attested by the depositions taken in
presence of the local authorities ; and that even a Scotch naturalist,
Sir Everard Home, proposed to class this monster among the fish of
the Squalidæ family, under the name of *Squalus maximus*.

But let us put aside these fables, legends, nocturnal visions, and
apocryphal narratives, and see what contemporary history, the reports
of men reputed worthy of belief, and the discussions of scientific
authorities can teach us in reference to this problematical being, whose
existence has sometimes been surrounded with absurd mystification,
sometimes asserted as an indisputable fact, while, until an epoch very
near our own, it was impossible to decide with certainty between
these contrary opinions.

In England and the United States a belief in the great sea-serpent
is (or was) exceedingly popular. The Linnean Society of Boston
published some years ago an authentic report establishing the fact
that, at certain intervals, a prodigious monster had been seen in
Boston Bay ; that on one occasion it showed itself, in 1817, about
thirty miles from Boston, and was examined by some competent
persons informed of its return. According to the narrative we are
speaking of, the monster exhibited the general shape and outlines of
a serpent. Its agility was extreme. When the weather was calm
and the sun hot, it remained on the surface, alternately plunging in
the water and exposing in the air the different portions of its an-
nular body.

In the archives of the town of Plymouth is preserved a long
abstract of verbal depositions made by a multitude of seamen, which
all affirm the existence in ocean of this mysterious animal. And it
is a remarkable circumstance that all these depositions, with the
exception of some slight differences of detail, fully agree upon the
general conformation and enormous dimensions of the monster.

A fisherman attests upon oath to have seen a strange animal, of
a serpent's shape, extraordinary size, and brown hue, sometimes
basking tranquilly on the surface of the water, sometimes swimming
with incredible swiftness. Another witness affirms that he saw in the

same locality an immense beast, whose head, said he, resembled that of a rattlesnake. A third had seen the monster open its enormous mouth, which he also compares to that of a terrestrial serpent.

Other individuals announce analogous facts, and accompany them with details which appear very natural. Thus, a seaman relates that he fired a musket-shot at the monster, just at the moment that, having drawn tolerably near the ship, he dived as if to avoid it ; but that, at a short distance off, the monster raised its head anew ; that they simultaneously felt the grating of a scaly body against the vessel's keel, and that soon afterwards they saw the serpent's tail lashing the surface of the sea, and making the spray and foam besprinkle the very mariners.

In the month of August 1819 the *United Service Journal* inserted a letter in which an eye-witness described the appearance of the sea-serpent on the shore of Nahant. " I had with me," says this witness, " an excellent telescope. When I reached the strand I found many persons assembled, and soon afterwards we saw appear, at a short distance from the shore, an animal whose body formed a series of blackish curves, of which I counted thirteen. Other persons estimated the number at fifteen. The monster passed thrice at a moderate speed, traversing the bay, whose waters writhed in foam under its huge bulk. We could easily calculate that its length could not be much less than fifty to sixty feet. . . This, at least, I can affirm, without presuming to say to what species belongs the animal which I have just seen, that at least it was neither a whale, nor a cachalot, nor any strong souffleur, nor any other enormous cetacean. None of those gigantic animals have such an undulating back."

A short time afterwards the officials of Essex county, in the State of Massachusetts, received the deposition, formally drawn up, which follows :—

" I, the undersigned, Gresham Bennett, second master, declare that on the 6th of June, at seven A.M., while navigating on board the sloop *Concord*, on her way from New York to Salem, the vessel being about fifteen miles from Race Point, in sight of Cape St. Anne, I heard the helmsman cry out, and call me, saying that there was something close to the ship well worth looking at. I ran immediately to the side which he pointed out, and saw a serpent of enormous magnitude floating on the water. Its head rose about seven feet above the surface; the weather was clear and the sea calm. The colour of the animal in all its visible parts was black, and the skin appeared smooth and free from scales. Its head was about as long as that of a horse, but was the perfect head of a serpent, terminating on the upper part in a flattened surface. We could not distinguish its eyes. I saw it clearly from seven to eight minutes; it swam in the same direction as the sloop, and went nearly as quickly. Its back consisted of humps or rings of the size of a large barrel, separated by intervals of about three feet. These rings appeared fixed, and resembled a chain of hogsheads fastened together; the tail was beneath the water. The part of the animal which I actually saw measured about fifteen feet in length; the movement of its rings seemed undulatory."

Thenceforth, and down to an epoch very near the present time, not a year passed but the presence of the sea-serpent was remarked at some point of the ocean. But the public speedily grew weary of these stories, and the great majority of cultivated minds saw in their authors only visionaries or mystificators.

However, in 1857 the question of the sea-serpent was again brought before the world by an English seaman of recognized ability, Captain Harrington, commanding the ship *Castilian*. There ensued in the scientific journals and societies, especially at London, a very animated discussion, but one of novel character, in which everybody took a side for or against the great sea-serpent; only its opponents, instead of denying purely and simply its existence, maintained that what had been taken for an animal was nothing else than some enormous vegetable waif.

But we must not anticipate, and will let the observers speak for themselves.

Captain Harrington professed to have seen the serpent very dis-

TO BE, OR NOT TO BE?

tinctly. According to him, the monster's head was shaped like a barrel, whose major diameter would be between two and three feet. On the top of the head rose a sort of membranous and wrinkled crest. For upwards of a hundred feet around the animal, the sea was agitated and discoloured, so that the captain's first impression was that his ship had got into what the sailors call "broken waters," and which are attributed to some submarine volcanic phenomenon. But a closer examination convinced him that before his eyes was a living being, of extraordinary length, apparently directing its slow course towards the land. At the time his ship was sailing too swiftly for him to measure the animal's dimensions, but according to such calculations as he was able to make, it appeared to be more than two hundred feet long. " I am convinced," added Captain Harrington, " that this animal belonged to the serpent species ; it was of a sombre colour, and covered with white spots."

The narrative in all its details was clear and precise. The captain wrote boldly to the Admiralty that, as a seaman, he could not be deceived, and that he should be as capable of mistaking an eel for a whale, as algæ or any other marine production for a living animal. " If it had been some distance off," he said in conclusion, "I should have thought myself mistaken; but I saw it pass within twenty yards of my vessel. A score of persons saw it as well as myself and my two officers, and I can assure you that I saw it as distinctly as I see at this moment the jet of gas in whose light I write to you this description."

In the face of statements so distinct and so categorical, the most incredulous hesitated ; many avowed themselves convinced ; and the cause of the sea-serpent was nearly won, when a new champion suddenly appeared in the arena. This was another sailor, Mr. Frederic Smith, who came forward as an eye-witness of its *non-existence* !

In the month of December 1848, Mr. Smith was sailing on board the ship *Peking*, belonging to his father, when, near Moulmein, in calm weather, he saw at a certain distance "something extraordinary

THE SEA-SERPENT.

balancing itself on the waves, and which appeared to be an animal of immeasurable length. With our telescopes," he added, "we could from the *Peking* perfectly distinguish an enormous head, and a neck of monstrous size, covered with a mane, which alternately appeared and disappeared. This appearance was likewise seen by all our crew,

LEGENDARY SEA-SERPENT.

and everybody agreed that it must be the great serpent. I took the resolution of making the closer acquaintance of this celebrated monster, and immediately ordered a boat to be lowered, with an officer and four men on board, furnished with some arms and a few fathoms of rope. I watched them attentively. The monster did not

A MONSTROUS ALGA.

seem disturbed by their approach. At length they arrived quite close to its head. They appeared to me to hesitate ; then I saw them busily unrolling the rope with which they were provided, while the monster still continued to raise its head and unfold its enormous length. Suddenly the boat began her course to regain the vessel, followed by the formidable monster. In less than half an hour the latter was hauled on board. The body appeared endowed with a certain suppleness so long as it remained suspended. But it was so covered with marine parasites of every species, that it was not until some time had elapsed we arrived at the discovery that this terrible animal was neither more nor less than a monstrous *alga*, upwards of one hundred feet long and four feet in diameter, whose root at a distance had represented its head, while the motion communicated to it by the waves had given it the semblance of life.

"In a few days this curious alga, growing dry, spread through the ship so infectious an odour that I was compelled to have it cast into the sea. Immediately after my arrival in London, the *Dædalus* reported its encounter with the great serpent in nearly the same parts, and I cannot doubt but that it was only the floating wreck of the alga whose history I have just related. Nevertheless, this illusion is rendered so justifiable by the appearance of the object, that if I had been unable to despatch the boat at the moment as I did do, I should have remained all my life in the conviction that I had seen the great serpent of the sea."

This relation stands in no need of comment ; it definitively settles the question, explaining by the most natural fact in the world the errors of all those who pretended to have seen the sea-serpent, but had only seen it at a distance, and had never dared, like Mr. Smith, to seize it by the body. Mr. Smith clearly accounts for the illusion of which his comrades had been the dupes, and which he himself, as well as his crew, experienced. It is certain that the aspect and atmosphere of ocean singularly predispose the mind to nourish hallucinations. No better proof of a fact so astonishing and so dramatic is needed than that in which M. Felix Julien was an actor, and which

20

I have related in a preceding chapter. We may readily understand, therefore, how, under this peculiar influence, which into the dullest mind infuses some slight feeling for the romantic and some slight sympathy with the wonderful, the gravest and most cultivated seamen are unconsciously terrified by the sudden appearance of the trunks of algæ of the fantastic kind described by Forster, and whose immense stem, undulating on the surface of the waves, easily deceives the imagination into mistaking it for the sinuous creeping body of a gigantic reptile. [Something, moreover, is due to the influence of ancient traditions and venerable fables, which have been handed down from generation to generation, and which, while powerfully affecting the more credulous and impressible minds, are not without their effect even upon cooler judgments. The superstitions of the past have a strange vitality in them. We pretend to despise, to ignore them; we very learnedly discuss their origin, and expose their absurdity; yet who can say that he is wholly free from their far-reaching power? Unknown to ourselves, perhaps, they colour our fancies and direct the course of our thoughts, and surprise us into a sudden acquiescence in moments when the cool intellect is off its guard, and the excited brain has surrendered itself to the dominion of Fancy. It is to this truth Schiller has so finely alluded in his "Wallenstein," in a passage where Coleridge's translation may be owned to surpass the original:—

> . . . . " Still
> Doth the old instinct bring back the old names;
> And to you starry world they now are gone,
> Spirits or gods, that used to share this earth
> With man as with their friend;
> Yonder they move, from yonder visible sky
> Shoot influence down; and even at this day
> 'Tis Jupiter who brings whate'er is great,
> And Venus who brings everything that's fair."

Thus influenced by the traditions of the past, the mind is prepared to be easily deceived; and the seaman, taught from his childhood to credit the existence of the terrible ocean-monster, readily gives its form and semblance to a mass of floating weed.]

MONSTROUS SEA-WEED (ALGA) DISCOVERED BY THE BOAT OF THE BRITISH SHIP "PEKING."

Who knows but that some day the problem of the gigantic polypus will be resolved in the same way as that of the sea-serpent! Thus, between the history of the former and that of the latter, will be a last feature of resemblance which should not at all surprise us. Since so many enlightened observers have affirmed, in all good faith, that they had seen with their own eyes the great oceanic reptile, we may well suppose that others have been equally mistaken in reference to the gigantic polypus—have been deceived by the trunk of some monstrous fucus uprooted from the bed of the sea, and whose roots or branches, agitated by the waves, imitated and resembled the tentacula of a cephalopod. It is also very possible that the fragments collected by certain voyagers, and represented by them as portions of polypi or enormous calamaries, were, in reality, the *débris* of a marine plant. The soft consistency of these fragments, their brown or reddish colour, their viscous surface, and the strong odour which they exhale, are so many characteristics equally proper to a great number of the products of ocean, and which we have no reason to attribute to an animal in preference to a vegetable substance.

These, and other considerations which a little reflection will suggest to the reader as to myself, should suffice, I think, to induce all sober minds, and especially men of science, to discard as exaggerations the narratives I have recorded of those extraordinary beings whose existence would, in a certain degree, be a negation of the great laws of equilibrium and harmony which govern living nature no less than brute and inert matter.

LOCH NESS

...To get the right answers, you have to ask the right questions...

by

Martin F. Jenkins

(I drafted this article in Liguria late on a May evening, with a glass or two of Dolcetto d'Alba 1995 in front of me and a chorus of frogs in the hills behind me. One or the other may explain my conclusions - or not, as the case may be.)

I am not certain of many things; but among my certainties is that we will never see a living dinosaur. All our understanding of evolution and ecology is against it. If a dinosaur population and its ecological niche had survived the great extinction sixty-five million years ago, the dinosaurs would have had an immediate advantage. They merely had to breed to refill the niche; the mammals had to evolve and breed to fill it.

If, however, I am wrong, I hope that when a relict dinosaur is found it is of a totally unknown species. Bernard Heuvelmans' seminal book was entitled *On the Track of Unknown Animals*. Too many contemporary cryptozoologists could write a book called *On the Track of Very Well-Known Animals that we hope are not extinct*. Nothing is more academic, that is, more un-Fortean, than the attempts to squeeze reports of mystery animals into known categories. It maintains the zoological world within safe, known boundaries. The coelacanth does not threaten the academic world-picture: we knew it was there once, we were rather surprised to find it still around. The okapi - an animal that everyone failed to predict - is another story.

After that opening, no-one will be surprised to learn that I do not believe the Loch Ness monster to be a plesiosaur. You may also have guessed that I don't want it to be a plesiosaur - and you will read the rest of what I have to say with that bias in mind.

If any creature survived the great extinction, the plesiosaur would be a good candidate. The ocean is a relatively more stable environment than the land, and if any part of it becomes unstable or uninhabitable it is easy to swim to other parts of it. Even after the catastrophe, whatever it was, there was still plenty of fish in the sea; the plesiosaurs need not have gone hungry. The only modern marine carnivores of equivalent size are sharks, giant squid and whales (and the last evolved from later land mammals and were not around sixty-five million years ago). If a breeding population of plesiosaurs had survived, the seas would be swarming with them today (unless humans had hunted them to extinction).

However, the plesiosaur concept of Nessie has so taken hold that it provides the basis for a classic chain of scientific illogic:
~ the Loch Ness monster is a carnivorous plesiosaur and eats fish
~ but there are not enough fish in Loch Ness to sustain a population of large carnivores
~ therefore, the Loch Ness monster does not exist.

Unfortunately, some cryptozoologists are lured by this into arguing about how many fish there are in Loch Ness, instead of pointing out that the third term should read: "therefore, the Loch Ness monster, if it exists, is not a plesiosaur or any other carnivore."

If we approach the evidence with an open mind, there is enough information contained in the various sightings for us to form an idea of what kind of unknown animal Nessie is. At this stage in the evolution of life, any new species of large animal is going to be related to something else in the animal kingdom; it will fall within some broad known category. (This assumes that Nessie is a physical creature, and not a psychic phenomenon - or a stranded extraterrestrial which keeps popping its head out of the water hoping for the rescue ship.) By analysing the key features of the sightings we can get a good idea of that broad category and of how to proceed with looking for the monster.

I want to stress three key features:
(1) the monster has often been seen in the Loch semi-submerged or travelling just under the water, more rarely with its head out of water

(2)   the monster has been seen on land

(3)   sonar sweeps have failed to find the monster swimming in the loch.

Now, the first of those observations suggests that the monster is not only basically aquatic but also that it derives its oxygen from water. An air-breather would have been seen above the surface far more frequently, because it would have to surface quite often to breathe, and one would expect reports of whale-like spouting.

However, the land sightings suggest that it is also able to breathe air. I am not convinced by Neil Arnold's argument (*CFZ Yearbook 1996*, pages 127-131) that Nessie (or one of the Nessies) could be a giant eel. Eels, as he says, "have been known to take to the land," but only in extreme circumstances - usually to escape from a home which is drying up in search of other water. The land sightings do not sound like eels, and I believe that we must look for a creature that can breathe in both air and water.

Neil Arnold's theory might explain the third key observation. Sonar is very good at detecting creatures swimming in relatively open water, as a carnivore would be likely to do. It is less likely to detect a giant eel feeding on the bottom of the loch. There could be a whole population of relatively shallow-bodied monsters sluggishly feeding in the ooze and sonar would not notice them unless it were adjusted to show small variations in the configurations of the loch bed. If you are looking for something which you believe behaves in one way and in fact it behaves in a totally different way, you have a very strong chance of not finding it.

So what are we looking for? I would suggest searching for a bottom-feeding vegetarian and/or plankton-eater. (Is there plankton in Loch Ness? It would be worth finding out.) Records of sheep-snatching suggest that it will eat meat, when it can get it, though it may be too slow to catch fish regularly. The giant panda similarly spends most of its day eating bamboo - a very inefficient food source - but will eat meat left over from another animal's kill. If Nessie has to spend the bulk of its time feeding, it would explain the rarity of sightings.

An aquatic creature also at home on land would suggest some kind of amphibian. There are two objections to this. The first is the recorded size of the monster. However, there is no reason why an amphibian, like any other vertebrate, should not grow to that size, and there were large amphibians before the age of reptiles. The stronger argument is that a body of water as cold as Loch Ness is not really amphibian territory. However, this is only a subset of a larger argument: Loch Ness is not the natural home for any large cold-blooded animal (including a plesiosaur).

To get round that difficulty, one could postulate a water-breathing mammal: which would really raise a few cryptozoological eyebrows. It would still be a bottom-feeder, however; so the next step in any case may be to recalibrate the sonar to look for odd moving bumps on the bottom of the loch.

Whatever is found, I hope it surprises all of us - including me.

Loch Ness Diary 1997
Nick Molloy

Unlike the 1996 expedition, when I spent three nights sleeping rough on the shores of Loch Ness in search of the lake's mysterious denizens, this year I went prepared with some rudimentary basics. My first year in employment after graduation had permitted a more kitted expedition. For example, this year I was equipped with a tent. Not a good one, but a tent nevertheless. More importantly, after several weeks of protracted negotiations I had managed to persuade my mother to lend me her 12X zoom video camera. My recently purchased 100 litre rucksack also had ample room for my somewhat bulky binoculars. Of course, I also made sure there was room for a standard camera. Oh, and lest I forget, there was the girlfriend (feel for her), also equipped with a rucksack; The contents of which consisted almost exclusively of fruesli bars and blocks of cheese. Armed with a credit card in case of emergencies we boarded the train in London...............

Sunday June 29th

A ten hour train journey meant we arrived in Inverness just after eight in the evening. The taxi dropped us of at Lochend around about 21.30. It had been a gorgeous day and I thought to myself that we would be in for a good week of monster hunting weather (good job I'm not a betting man). We only walked for about a mile before we located an ideal camping site, right on the shore. Although the pebbled shore was a little uncomfortable to lie on, I explained to the other half that it was imperative to have an open view of the water at ALL times. She looked on a little incredulously.

I had been up all night on Saturday (boxing from Las Vegas), so I was fairly tired. It was still reasonably light at midnight, not long after which I nodded off.

I awoke at approxiamtely 3 am to the oddest animal sound I have ever heard. I instinctively scrambled for the camera and fumbled with the zip of the tent. Preparing to meet my doom, I emerged outside determined to leave as my legacy that crucial piece of evidence : the definitive photograph.

Opposite Urquhart

All photographs © copyright Nick Molloy / CFZ

My tired eyes struggled to adjust to the night/early morning gloom. I stood there for a minute or so and then the shriek echoed out again. It seemed to come from the mountains at our back. I would describe the noise as a demented dog/duck amalgamation. After subsequent discussions with locals the most likely candidate for the owner of the sound appears to have been a roe deer. A false alarm.

The loch surface was like a mill pond, it glistened invitingly in the moonlight. If I wasn't so tired from the previous night I would have lit a fire and begun my vigil. This would have meant sleeping during the day. Not a problem, except to the girlfriend (well I have to give a bit). I crawled back into the tent and went to sleep

Monday, June 30th

I arose about eight o'clock with the weather turning for worse. The inoffensive mill pond was turning choppy. Ominous looking clouds had now gathered overhead blocking out the sun. After a breakfast of Fruesli bars, the tent was packed and we began to proceed along the rocky shore (view imperative I reiterated). Anyway, carcass finding goes from a long shot to a non shot walking along the road I stressed. Scornful looks persisted for about a mile of shore until we arrived at a scree slope. We can scramble over that, I said. Fall in ? Never, I said. With the strong possibility of a separation looming, I was forced to scramble up onto the road.

The remainder of the morning was spent walking along the road towards Drumnadrochit. My beloved water view was all but obscured except for fleeting glimpses that arose through a gap in the trees and undergrowth. Arrival in Drumnadrochit meant a hot meal and a tour around the Official Loch Ness Exhibition. Although reasonably well presented, the exhibition's primary focus was the tourist, and will do little for serious Nessie students.

Leaving Drumnadrochit, in the late afternoon, we made our way up the winding road to Castle Urquhart and come within sight of the Loch again. As we reached the road overlooking the famous landmark, the rain began to come down. Rather than stopping, we decided to press on and look for a suitable camping area. With it becoming apparent that we would not find a suitable Lochside spot we decided to keep walking, into the night if necessary.

Just woken up!

The mist approaches

Pine Marten (dead centre)

Opposite Urquhart

With the light fading fast and only about a mile short of Invermoriston, it *really* began to rain. A strong wind began to pick up. Fearing a soaking, we hurriedly set up tent in a field just off the road and scrambled inside.

What followed almost led to an early return home. A violent storm ensued. Torrential rain and very strong winds almost caused the tent to lift off ! I subsequently found out, next, day, that the nearby town of Moray was under a few feet of water. An American tourist later informed me that the River Ness was also very close to bursting forth into the streets of Inverness on Tuesday.

Tuesday, July 1st

We awoke inside a tent/swimming pool. Everything was soaking; sleeping bags, clothes, everything. You can guess my level of popularity at this point.

Everything was packed away, wet, and the short walk into Invermoriston followed. Over a hot breakfast a journey home was averted and it was decided that a bus would take us to Fort Augustus where at a bed & breakfast we would try and dry out equipment. Fortunately, our bed and breakfast had an electric fire allowing us to dry most of it.

Despite the home comforts I suggested that it would still be better if we were camping out on the shore...... watching. My case was not helped by the fact that it was still beating down with rain outside and I suppose I should have been prepared for the stern rebuffal.

Disasterously, the night before had been more than just uncomfortable. The main battery and the backup battery for the video camera had got damp. The camera was without power and the charger, being bulky, I had left at home. My Tim Dinsdalian efforts had been thwarted before they were barely under way. Furthermore, water had seeped into the eye of one of the binocular lenses, leaving me vision out of one eye only (looking on the bright side, at least I could use my good eye)!

As rain was forecast for the next night it was decided that we would again stay indoors the next night. Foyers was the next destination on our walk round. The Foyers Hotel was the obvious choice. After all, had the great man himself not stayed there on his historic first expedition. If I could not know emulate Dinsdale in recording a historic piece of film,

Boleskine House

'Deserted side' near Lochend

I could at least follow in some of his footsteps.

However, protests were raised about the price of the Foyers Hotel establishment in comparison to that of others. "Dinsdale, Dinsdale, hail Dinsdale" was my original opening. It degenerated into who's credit card is it anyway.

Wednesday, July 2nd

An early rise meant a Loch Ness cruise aboard the Royal Scot, before our journey to Foyers. I spent the first hour half, eyes fimly fixated on the water and paid little attention to the ramblings of our driver/guide. I spotted one of Steve Feltham's small Nessie model's inside the cockpit area and said so to the other half. Clearly surprised that I knew who Steve Feltham was, the guide stated that he and Steve were now the only two Nessie researchers permanently based at the Loch.

This would have been quite interesting, but for the utter garbage our driver/guide then began to spew forth. His echo sounding equipment has apparently detected some items of great interest. These include 30 foot eels. Yes 30 foot. An 18 foot pike. Yes, that is bigger than the average great white shark. Also, he has apparently been monitoring a nine foot baby Nessie born in 1994.

As we were pulling back into port I asked him what he thought of the photographic evidence for Nessie. He stated that he only believed that one was genuine. "R. K. Wilson's photo" he proudly stated. "Ah, the Surgeon's photo" I replied. A model mounted on a submarine I suggested. "Clockwork submarines weren't invented then" came back the reply. After which he declined to discuss anything further, saying he had to hurry up and guide the next trip.

For those gullible enough to believe any of the above, why not sponsor a Loch Ness Plesiosaur. For only £20, you too can part own your very own Nessiteras Rhombopteryx. Cheques payable to N. Molloy, care of the editorial address.

A taxi took us halfway to Foyers allowing an arrival at the Foyers Hotel in the late afternoon. The girlfriend was right, it was twice the money, for only half the facilities, but, it did have that all important Loch view. I explained that this was worth the extra investment. Something abusive was uttered in my direction.

I spent half an hour scanning the Loch from our bedroom window, using the now one eyed binoculars, before I was ordered out to buy food. On our return, about 18.00, I again returned to the window. Almost directly opposite, on the far side of the loch, perhaps three to four hundred yards off the far shore, I could make out a dark green spec on the water's surface. Scrambling for the binoculars, I focused in on the area.

Through the binoculars, I could make out the shape of the object. A tall object, wider at the base, almost like an elongated triangle. Watching it for nearly a minute, it had not moved and I concluded that it was in fact some sort of water buoy. With my binoculars in a weakened state I could not be sure of this conclusion however.

Confident that I was not looking at anything exceptional I scanned with the binoculars elsewhere on the loch surface. Approximately, 30 seconds later, I returned to my buoy. It was still there unmoving, unblinking. I was now convinced that it was an inanimate, lifeless object. Again I began to scan other parts of the loch. A small boat was visible perhaps a mile to the west of my buoy. Again, 30 seconds later I returned to my buoy, in order to cement my conclusion. Yet, my 'buoy' was no longer anywhere to be seen. I scanned with the naked eye and then with the binoculars. The buoy had vanished.

The explanation for this incident is quite probably obtuse and mundane. However, I don't have one ! Something was there and then it wasn't. No boat could outrun the reach of my vision in such a short space of time. Furthermore, I would estimate the height of the object at four to five feet based on the boat I saw to the West of the 'buoy'. I'm not suggesting I saw the Loch Ness Monster, but, I did see something unusual and as yet unexplained. Suggestions gratefully received.

That evening I left the girlfriend at the hotel watching TV and climbed down a cliff track into Foyers Bay. Perched on the shore I remained for about an hour, watching. At one point I thought that a wake was being created mid loch by an underwater force. Sprinting back for a higher vantage point however, it became obvious that it was instead one of those strange Loch Ness surface effects. The effect continued to ripple northwards on a glassy surface, creating another illusionary sub-surface monster. As I made my way back to the hotel, I began to wonder where the forecasted rain had got to ?

River Ness

Dores

After breakfast, there was one last bedroom vigil and we again took to the road. We continued past Boleskine Burial Ground with its eerie Loch view, past 'The Wall' and found an ideal camp site, in a copse, off the road, with an excellent Loch view, opposite Urquhart bay. The day passed uneventfully, apart from the complaints about midges and the taste of Fruesli bars. I went to sleep at midnight and arose at dawn to begin a vigil. The weather however was far from welcoming. A chilling wind did not encourage the fire, or me for that matter. The loch was very choppy and conditions were therefore far from ideal.

After about 20 minutes, I abandoned the futility of the fire and returned to the tent. The tent was now so useless, even the morning dew was soaking through. I drifted back to sleep.

Friday, July 4th

We set off along the road to Dores at about 8am. After about another mile along the road we came to one of the many picnic areas on 'the deserted side' and clambered down to the shore. It was time to put my foot down and insist on a shoreline walk ! I won (shock) ! It took about an hour to go a mile. Climbing over tree trunks and under vines, whilst the rocky underfoot ensured a difficult trek. Nevertheless, I had an open water view and I did find an interesting bone. I'm guessing that it was once part of a deer's foreleg, but, remains unexplained.

After approximately an hour and a half of this very interesting and eventful shoreline trip, a separation was again pending, and I was forced to relent and return to the road. It was on the return to the road that two Pine Martens were spotted in a nearby tree. They r n upon seeing us, but, with patience, one of them again made itself visible and available for a photograph.

Dores was reached by late afternoon, where I hoped to meet Steve Feltham. Although, not in his motorhome when I arrived, he soon returned. I went over and introduced myself whilst the girlfriend shyly retreated. A nice chap, Steve introduced me to Jean Jacques Colin, a French artist on his seventh visit to the Loch who very kindly gave me one of his

View of 'sighting area'

prints. The print was an artists impression of Rines' underwater Gargoyle Head photograph (an excellent transformation of what is now widely accepted as a tree trunk).

Steve was adamant that he would be there until he solved the mystery of the Loch and then would be interested in going to Okanagan in search of Ogopogo. Bidding them farewell, we proceeded into the woods along the alleged Loch Ness trail. Finding a suitable place on the shore we set up camp. The night and dawn was much like the last and largely unsuitable for vigils. Frustrated again, I slept until 6am.

Saturday 5th July

We packed up at 6am and continued along the alleged Loch Ness Trail. Through the wood and along the River Ness. As a migration theorist, I believe that watching the River Ness could be just as important as watching the Loch itself. The river provides a suitable outlet from the Loch to the sea. When in flood such as earlier in the week, it does not seem unreasonable for the river to hide the passage of large aquatic creature/s to and from the open sea.

We progessed along the river bank as far as we could, before our course forced us back up onto the road. We decided on a bus into Inverness followed by a bus back out to Drumnadrochit. I was curious regarding the Official Loch Ness Monster Exhibition. This proved to be a disappointment. The highlight was a twenty minute film that looked and felt more than a little dated. A serious of boards displayed various Nessie photographs. I was dismayed to note that numerous Frank Searle photographs were being displayed as genuine evidence of the Loch Ness Monster.

There were however, some photographs of which I was not aware. However, in my opinion, they were obvious fakes. Lachlan Stuart's photograph sprung to mind. Tyres or bales of hay covered in tarpaulin, again ? I refer you to the sponsoring of a Loch Ness Plesiosaur mentioned earlier.

The final night was spent at the same site as our first night. With the weather having frustrated me from day one, I was determined to spend a proper vigil on the last night. I arose at 3am. It was blowing a gale outside, tsunamis were breaking on the shore with depressing force and regularity. I couldn't get the firelighters to light, let alone maintain a fire. Freezing, I again retired defeated by the elements.

Sunday, July 6th

Arising at 8am we walked back to Lochend and ordered a taxi in order to return to Inverness. The weather on the journey back was not unlike that on the journey up. I began to think that I was cursed.

In summary the fortean/zoological enigma of Loch Ness had eluded me. Although, I had a truely surreal alchemical experience whilst at the Foyers Hotel. My girlfriend was accusing me of various atrocities to which my only defence was Dinsdale. I had no idea if my mum's video camera would ever work again. I was covered in insect bites and would probably die of malaria. It's a monster hunter's life. Anyone for expedition '98 ?

Monday 7th July

London Commuting, AAARRRRGGGGGGHHHHHHH!!!!!!!!!

Nick Molloy on his previous expedition, 1996

Did a Comet Kill the Mammoths?

by

Emmet J. Sweeney M.A.

The latest revelations concerning the catastrophic demise of the dinosaurs can only make one wonder how long it will be before mainstream scholarship begins to take seriously the equally compelling evidence for a much more recent cataclysmic upheaval of nature - one which occurred just a few thousand years ago and is recorded in the folk memory and traditions of mankind.

I can only suspect that the sudden death of the dinosaurs, which is believed to have occurred 60 million years ago, can be freely discussed because of its very remoteness, and does not elicit the same anxiety as a similar event placed just a few thousand years ago would.

Yet the evidence for some form of giant upheaval of nature during the Stone Age is, if anything, even more compelling than the evidence for a similar event in the time of the dinosaurs.

Just as at the end of the Cretaceous period, there was a sudden mass extinction at the end of the Pleistocene. Hundreds of species, well adapted for survival in their native environments, disappeared forever. Best-known of these perhaps are the mammoth, woolly rhino, and sabre-toothed tiger. Huge herds of these and other extinct animals wandered throughout North America, Europe, and Asia. Their disappearance, when discussed at all, is normally explained as being gradual, and is linked to the onset of the last Ice Age. Yet such an explanation flies in the face of a number of very well-known facts, which are nevertheless studiously omitted from most of the geology and natural history textbooks.

Throughout many regions of the north, in Siberia, Alaska, and Canada, the bodies of vast numbers of animals, of species both extinct and surviving, are found frozen in the permafrost, just a few feet beneath the surface. These creatures are often perfectly preserved, with flesh and fur intact, though most have almost every bone in their bodies broken. Whole hecatombs of such beasts are found in wildly chaotic circumstances. Animals which would never associate with each other in life, carnivores and herbivores, are found heaped on top of each other, and their bodies intermingled with uprooted and smashed trees, along with other types of debris.

In some areas, most famously the muck-flats of Alaska, these deposits can be up to 50 metres in depth. Indeed, some of the islands of the Arctic Ocean seem to be composed almost entirely of this material.

So plentiful were the visible remains of mammoths in Siberia during the last century that a flourishing trade in mammoth ivory developed; and in fact much of the ivory used in Europe 100 years ago for piano keys etc. came not from Africa, but from the 'Ivory Mines' of northern Russia.

Some of the mammoths discovered in Siberia are so well preserved that their flesh can still be eaten, whilst a number have been shown to have traces of undigested buttercups and other flowering plants in their stomachs.

This is a fact of astounding importance, and its failure to be taken into consideration is an outstanding omission on the part of natural historians. These animals were evidently grazing in a meadow, on a warm summer's day, when their bodies were transported, with great speed and violence, into the Polar regions, and there deposited along with vast numbers of other animals, uprooted trees, and debris of every kind.

The early European discoverers of these creatures were in no doubt that one force, and one force only, could account for what they found. Giant tidal waves, named 'waves of translation' were the only thing that could uproot whole herds of animals, as well as forests, and leave them within hours in the permafrost regions.

It would take too long in a newspaper article to explain how the evidence from the Arctic was systematically ignored and eventually suppressed during the early years of the present century. Suffice to say that the new Darwinian orthodoxy was to decree the concept of catastrophism as incompatible with the idea of evolution (though interestingly Darwin himself initially had no problem with the idea, and during his voyage on the Beagle noted much evidence for violent cataclysms of the past along the coast of South America).

In vain will the modern student of geology or palaeontology search the textbooks for an honest appraisal of the mammoth problem. It is either ignored or brushed aside in a few pat sentences. Nevertheless, a number of scholars, both here, in Europe, and in North America, have swum bravely against the tide for many years, and it looks as if they are at last being heard. In recent years two prominent British astronomers, Victor Clube and Bill Napier, have published a series of successful books on the subject, whilst in Germany Dr Heribert Illig and Professor Gunnar Heinsohn, of Bremen University, are winning over many high-powered supporters.

But if modern man tries to ignore the events which deposited these creatures in the permafrost, ancient man, who witnessed them, did not.

The ancients told of a comet, variously described as the Cosmic Serpent, the Dragon Monster, etc., which had come close to the earth and caused a great flood. The Roman writer Pliny, who had access to much ancient material that is now lost, named the comet Typhon, and described it as blood-red with a great tail twisted like a coil. This heavenly visitor, he said, had brought the earth to the brink of destruction. Whole volumes could be filled from the traditions of every land about this cataclysm. One account, however, from the Incas of Peru, should serve to provide a fairly typical example of the genre.

It was said that in the remote time of the ancestors the world went dark for a space of nine days.

At the end of this period a great roaring sound was heard, and a light was seen in the distance. As the light grew closer however men saw that it was the crest of a gigantic wave.

How long will it be before such evidence is taken seriously, and the truth of our human and natural history (so much more exhilirating and dramatic than the history we find in the textbooks) is finally revealed? There were indeed, it would appear, many great cataclysms in the past, not one, and the last of these was witnessed by human beings, who built their mythologies and religions around it.

A Further Compilation

of Cryptozoological

Films

(and possibly the ultimate collection)

By Neil Arnold

Over a year ago, I suggested to Jon Downes that a list of films with cryptozoological themes should be compiled, and this became my contribution to the "1997 Yearbook". I have since explored the vaults of cryptozoological movies in greater depth, in search of hidden gems. Michael Playfair's compilation stimulated this sequel of my discoveries, which should provide comprehensive coverage of this topic. No other magazine has covered this aspect of the cinema in so much detail. The Bigfoot horrors, the serpent dramas, the Yeti B-films and the dinosaur adventures unfolded into a magical celebration of monsters on celluloid. It is not a difficult task to churn out countless "dinosaur" movies; which is why my original article, "Crypto on Celluloid," mainly covered true crypto-creatures like Nessie and Bigfoot.

I have now uncovered a vast number and variety of crypto-films; including some really obscure ones dating back many years. Of course, there are some crypto-movies that have been forgotten forever; endless dinosaur films and countless movies concerning over-sized creatures; and therefore no mention has been given to films like "Barracuda" and "Dogs". My last attempt to find a few more films has been very productive, and this will probably be the final list, for perhaps in another fifty years I might have enough new material to produce another list like this one.

" This is the greatest moment of my life, I've seen a living pterodactyl."
--- Professor Challenger
"The Lost World"(1925).

1) **The Wild Man of Borneo** (1902) Obscure movie, directed by Walter Haggar, which I know little about.

2) **Conquest of the Pole (1912)** Surreal tale by Georges Melies, concerning an expedition and a beast called the "Abominable Giant of the Snows".

3) **Gertie the Dinosaur (1914)** Comical short film about a captured dinosaur in a circus.

4) **Gertie on Tour (1917)** A sequel!

5) **Morpheus Mike (1917)** The first movie to feature a mammoth!

6) **The Dinosaur and the Missing Link (1917)** A tale of dinosaurs and missing links - what else?

7) **The Ghost of Slumber Mountain (1919)** Highly imaginative tale about a guy who meets a hermit. Through a viewing instrument he can see the mountain as it was in the past, alive with prehistoric beasts.

8) **Siegfried (1924)** One of the first tales of dinosaurs on the rampage in London.

9) **Creation (1930)** An unreleased film concerning a shipwrecked crew who stumble on a lost world.

10) **The Dragon Murder Case (1934)** A crypto-detective film. A body, covered in claw marks made by a mystery beast, is found in a swimming pool.

11) **The Beast of Borneo (1935)** A film full of jungle expeditions and evolutionary theory.

12) **Jaws of the Jungle (1936)** Forest natives are attacked by giant snakes and monstrous Devil Bats.

13) **Son of Ingaga (1940)** A giant ape searches for love. How nice!

14) **Valley of the Mists (1944)** Unreleased. A twelve year old boy raises a bull-calf, then trades it in for a legendary giant lizard, which lives in the hills. The beast is captured, but runs amok in the town.

15) **Return of the Ape-Man (1944)** This has nothing to do with the first "The Ape-Man" film; this concerns a man-beast revived after an expedition.

16) **White Gorilla (1945)** Another white ape adventure, similar to " White Pongo".

17) **Jungle Captive** (1945)   Two scientists find an ape-woman and indulge in some genetic engineering.

18) **Mr. Peabody and the Mermaid** (1948) A 50 year old man finds a wonderful mermaid and takes her home to meet the wife!!

19) **The Dragon of Pendragon Castle** (1950)   A children's fantasy about kids who find a sea serpent and keep it in their castle.

20) **Mark of the Gorilla** (1950) Jungle Jim searches for treasure, guarded by a giant ape.

21) **Killer Ape** (1950s-60s) Natives are attacked by a King Kong rip-off.

22) **The Jungle** (1950s-60s) Indian expedition to find prehistoric mastodons.

23) **Jungle Jim in the Forbidden Land** (1952) Early film for Weissmuller (Tarzan), where he tackles werewolf-type creatures, which form a missing link.

24) **The Neanderthal Man** (1953) A hairy man-beast terrorises women.

25) **Port Sinister** (1953) A submerged pirate city rises after an earthquake, but the treasure hunters find sea monsters instead of valuables.

26) **Road to Bali** (1953) This dive for sunken treasure includes comic encounters with a silly squid and a slap-happy gorilla.

27) **Gojira** (1954) Another giant creature emerges from the seas.

28) **Phantom from 10,000 Leagues** (1954) Another sea monster, this time a serpent ray.

29) **Sword of the Dragon** (1956) A Russian film about a warrior who slays Hydra and other monsters.

30) **Curucu, Beast of the Amazon** (1956)   A hideous beast of the Amazon commits grisly murders. Menacing Brazilian locations used.

31) **Rodan** (1956) An atomic bomb test reveals two eggs, which hatch out into two massive pterodactyls. The film also boasts giant insects.

"SHRIEK OF THE MUTILATED"
College students seeking The Abominable Snowman find a cult of killers on an island of terror
IFS
IVER FILM SERVICES
THE PROFESSIONALS AT PINEWOOD

32) **Moby Dick** (1956) The classic tale of the great white whale.

33) **The Mole People** (1956) Weird albino-type creatures are found in remote Asia.

34) **Ghost of Dragstrip Hollow** (1956) Awful teen musical features an aquatic monster.

35) **Animal World** (1956) Semi-documentary delves into anthropology and dinosaurs.

36) **The Cyclops** (1957) An expedition to Mexico to search for a missing man, finds a 50 foot beast instead.

37) **The Giant Claw** (1957) A huge thunderbird-type creature battles the world and tries to lay an egg on the Empire State Building.

38) **The Deadly Mantis** (1957) An over-sized praying mantis wreaks havoc in the city.

39) **The Black Scorpion** (1957) Another ridiculous sized thing on the loose.

40) **The Eagle** (1957) Unreleased. Concerns a rancher who nurses an eagle. Soon livestock is mangled and the eagle is blamed. The truth is, a dinosaur is on the loose.

41) **Last of the Labyrinthodons** (1957) Unreleased. A never-filmed story of modern day vessels being attacked by sea monsters.

42) **Panda and the Magic Serpent (1958)** A dragon God and a legendary immortal serpent perform in this Chinese fairy tale.

43) **Monster on the Campus (1958)** A scientist turns himself into a Neanderthal by using coelecanth serum.

44) **The Giant Behemoth (1959)** An enormous sea beast crushing everything in its way.

45) **Caltiki, the Immortal Monster (1959)** A legendary monster dwells in an Aztec temple and kills people.

46) **Darby O'Gill & the Little People** (1959) Sean Connery stars in this spooky tale, supernatural entities from Irish folklore include leprechauns and banshees.

47) **Green Mansions** (1959) Audrey Hepburn plays the "Bird-girl of the Amazon" .

48) **The Killer Shrews** (1960s) More over-sized terrors.

49) **The Giants of Thessaly** (1960) Imaginative adventure in which a warrior battles over-sized monsters.

50) **Goliath and the Dragon** (1960) Hercules battles with giant bats, three-headed dogs and a dragon.

51) **Monster of Highgate Ponds** (1960) Action combined with animation. Kids find an egg that hatches into a monster. The beast then leaves Britain for Malaya.

52) **Terror Beneath the Sea** (1960s) Obscure tale of an angry sea-serpent.

53) **The Trollenbourg Terror** (1960s) An ugly, tentacled thing terrorises the Swiss Alps.

54) **The Creature** (1960s) A BBC play written by Nigel Neale, which inspired the Hammer film "The Abominable Snowman"

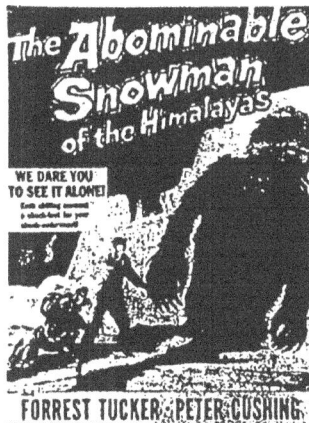

55) **Jack the Giant Killer** (1961) Jack tackles an octopedic sea monster and a griffin.

56) **Minotaur, Wild Beast of Crete** (1961) A legend of the mythical beast.

57) **Varan the Unbelievable** (1961) Another terrifying scaled serpent.

58) **Mermaids of Tiburon** (1962) Tale about a sexy mermaid and a treasure hunter.

59) The Blancheville Monster (1962)   A Spanish-Italian film about a maniac man-beast on the loose.

60)   Eeagh!! (1962) A very stupid movie about mumbling Neanderthals.

61) Horror of Party Beach (1963) Aquatic beasts slay the local townsfolk in this dated B-movie.

62) The Creeping Terror (1964)   Tacky film about a horrible lake monster in Nevada.

63) The Mighty Jungle (1964) Two explorers go on separate trips to the Amazon and the Congo, and discover pygmy tribes and iguanas.

64)   Mike and the Mermaid (1964) A boy tries to convince his elders that he has seen a mermaid in the river.

65)   Ape Woman (1964) A woman ape-beast is found and used in a freakshow.

66) Seven Faces of Dr. Lao (1964) Features a serpent and a Yeti.

67) Swamp of the Lost Monsters (1964) Lightweight Mexican movie. A woman is attacked by a swamp monster.

68) Gamera (1965) Apparently a "goodie", but still a massive turtle!

69)   Legend of Blood Mountain (1965) A newspaperman hunts a beast in darkest Georgia.

70) City under the Sea (1965) A land inhabited by a tribe of aqua-men.

71) Curse of the Swamp Creature (1966) In the Everglades, a swamp beast is spawned in order to kill off any stray persons.

72) Journey to the Beginning of Time (1966) Czech film mixing animation with real film as youths drift down the river of time and see dinosaurs.

73)   Night Tide (1967) Dennis Hopper becomes intrigued by a mermaid named Mora.

74) Yongary, Monster from the Deep (1967) A beast rises from the ocean after an atomic

blast.

75) **Gnome-mobile (1967)** Walt Disney musical about some children's encounter with elves.

76) **The Vulture (1967)** Turgid tale about a winged creature.

77) **War of the Gargantuas (1967)** A film featuring a giant octopus and others.

78) **Moonshine Mountain (1967)** A story about a killer ape.

79) **Gappa the Triphibian Monster (1967)** "Even mightier than King Kong" reads the poster. A hideous, beaked lizard.

80) **Sound of Horror (1967)** An oddity about a Greek expedition scared off by a prehistoric beast, never actually shown in the film.

81) **Scullduggery (1969)** Burt Reynolds joins a New Guinea expedition and discovers a tribe of ape-men, when all he was seeking was phosphorous!

82) **Shark (1969)** After "Jaws", these films fall flat.

83) **The Last Unicorn (1969)** An entertaining fantasy, which introduces us to the hideous Harpy.

84) **The Bear (1970)** A Polish folklore tale about a man-beast.

85) **Island Claws (1970s)** A completely unknown title, about a monster on a forgotten island.

86) **Monster Hunter (1970s)** Another obscure film about a monster hunter. You guessed it!

87) **Beast and the Vixens (1970s)** Very camp and funky film about Bigfoot and bimbos.

88) **Tanya's Island (1970s)** An erotic "Robinson Crusoe" type film. A stranded lady falls for a shaggy ape-man; which kills her.

89) **Saga of the Viking Women & their Voyage to the Waters of the Great Serpent (1970s)** Need I say more?

90) **The Night of the Sasquatch (1970s)** Backwoods folk are haunted by a Sasquatch. A

mixture of "Boggy Creek" and "Night of the Demon".

91) **Equinox (1971)**   An ancient witchcraft tome releases a pterodactyl-type monster after a disturbance.

92) **Long, Swift Sword of Siegfried (1971)**   A knight battles a legendary beast.

93)   **The Little Ark (1971)** Two children trapped in a church after a flood are saved by a fisherman who tells them about a legendary mermaid.

94) **Prelude to Taurus (1972)**   Arctic scientists discover frozen bodies over a million years old.

95) **Zeppelin vs Pterodactyls (1972)**   A Hammer film never made. It concerns your average "man versus dinosaur" theme.

96) **Horror Express (1972)**   Christopher Lee is the anthropologist who finds the fossilised missing link in China. Unfortunately, it comes to life on the way home.

97) **Frogs (1972)**   Inane over-sized creature film

98)   **Savages (1973)**   Primitive mud-people rise from the New York ooze.

99)   **The Neptune Disaster (1973)**   An underwater fantasy with mysterious sea-beasts.

100) Schlock (1973) Also known as The Banana Monster. A Neanderthal man-beast

101) The Sentry (1974) This was an episode in the "Nightstalker" series. Bumbling news reporter Carl Kolchack investigates a lizard man living underground, in the way of a construction project.

102) The Tribe (1974) A film of Neanderthals and Cro-Magnons.

103) The Spanish Moss Murders (1974) Another episode from the "Nightstalker" series. A legendary swamp creature terrorises the Cajun community of Chicago.

104) Primal Scream (1974) Yet another one from the "Nightstalker" series. A Chicago based oil corporation discovers frozen prehistoric cells in the Arctic. On thawing, they evolve into an ape-man.

105) Scream of the Wolf (1974) A Richard Matheson tale about weird beastly killings.

106) Matchmonedo (1974) A re-edited version of an episode from the "Nightstalker" series. This features a legendary beast that stalks around a hospital.

107) Island at the Top of the World (1974) A team on an expedition to the polar regions find a tribe of cannibal Vikings.

108) Man-beast: Myth or Monster (1974) Excellent documentary on Bigfoot.

109) An Encounter with the Unknown (1975) A horror trilogy. One of the tales has an unusual and out of place animal that lives in a hole.

110) Panic in the Wilderness (1975) A Bigfoot is on the hunt in the Canadian Northwest.

111) Lorelei's Grasp (1975) Spanish film about a murdering beast-woman.

112) Mako, Jaws of Death (1976) More over-sized rampant sharks.

113) In Search of Bigfoot (1976) .A docu-drama on the elusive Sasquatch.

114) Legend of Bigfoot (1976) Another Docu-drama in scarch of Bigfoot.

115) Squirm (1976) Excellent story about a plague of lethal worms. Better than most of the

"beast-on-the loose" type of drama.

116) A*P*E* (1976) A 36 foot high gorilla is found on a pacific island, captured, lost and pursued in Korea.

117) The Mighty Peking Man (1977) An expedition brings a giant ape out of the jungle and into the city.

118) Out of the Darkness (1977) Donald Pleasance pits his wits against a giant big cat.

119) Nessie (1977) Another Hammer film that never was. Nessie's escape from the loch was planned, with all the havoc of a typical sea monster drama. It had a budget of £3 million, but the idea faded.

120) Sasquatch (1977) No prizes for guessing what this is all about!!

121) Tintorera (1977) Another huge shark ripping off "Jaws".

122) Tarantulas: The Deadly Cargo (1977) An aircraft crashes; spilling thousands of lethal banana spiders.

123) The Shark's Cave (1978) An Italian "Jaws" imitation.

124) Planet of Dinosaurs (1978) More dinosaurs on the rampage.

125) The Bermuda Depths (1978) An expedition in search of a giant sea-turtle.

126) The World Beyond (1978) A muddy swamp monster terrorises the island locals.

127) Mystery of the Golden Eye (1978) Hypothetically true story. A senator is lost on an island populated by prehistoric dragons.

128) Gold of the Amazon Women (1979) A tribe of zebra-knickered women are found in the Amazon by lucky explorers.

129) Screams of Winter Night (1979) Horror anthology including a Bigfoot tale.

130) Up from the Depths (1979) More flesh-eating killer fish.

131) **Mistress of the Apes** (1979) Low budget film about the missing link.

132) **The Great Alligator** (1980s) No prizes for guessing the theme of this one!!

133) **Track of the Moon Beast** (1980s) A man-beast stalks and slaughters everyone.

134) **Terror in the Swamp** (1980s) Terror comes in the shape of a Sasquatch.

135) **Devil Fish** (1980s) Not quite as bad as "Piranha", but still as basic.

136) **The Strangeness** (1980s) A deserted goldmine reveals a sea beast known in American Indian tradition. The group of explorers are eliminated.

137) **Killer Fish** (1980s) A gang of robbers stash their diamonds in a lake infested with demonic fish.

138) **Monsteroid** (1980s) Fishing-folk are terrorised by a monster in the local lake.

139) **To Catch a Yeti** (1980s) Obscure fantasy movie that may have starred Meatloaf, and I think the Yeti is only pocket-sized.

140) **Blood Beach** (1980) Bathers are sucked into the sand by a kind of rubbery deathworm.

141) **The Ivory Ape** (1980) A large ape is captured and escapes to Bermuda. Soon Anthropologists are battling hunters in the search.

142) **Guardians of the Deep** (1981) More Italian killer sharks.

143) **Clash of the Titans** (1981) Everyone's favourite fantasy featuring the mighty Kraken and other mythological favourites.

144) **Night of the Rat** (1981) Tale taken from the "Nightmares" anthology; about a mythological rat-beast from Germany.

145) **Quest for Fire** (1982) Gritty film about Neanderthals; plus good footage of mammoths and sabre-toothed tigers.

146) **Nightbeast** (1982) Appalling film. A rubbery beast lays into its prey.

147) L' Ultimo Squalo (1982)   Another Italian Great White Shark movie.

148) Humungous (1982) Poor slasher film about a wild man killing youths.

149) Attack of the Beast Creatures (1983) Shipwreck survivors are washed up in the North Atlantic and attacked by a lost tribe of beastly warriors.

150) Luggage of the Gods (1983)   Caveman tribes come into contact with the present day.

151) What Waits Below (1983) Installation procedures in an underground cavern are halted by a huge monster and a tribe of albino weirdoes.

152) Dance of the Dwarves (1983) An expedition to the Philippines reveals a tribe of reptile-men.

153) Monster Shark (1984) A hideous shark-octopus kills swimmers.

154) Splash (1984) Crap comedy stars Daryll Hannah as a mermaid.

155) Terror in the Swamp (1984)   A giant swamp rat is killing off the hillbilly folk.

156) Encounters in the Deep (1985) Set in the Bermuda Triangle with an array of sea monsters.

157) Zoo Ship (1985)   An original tale about a spaceship, which crashes, carrying all the creatures of the universe. And we wonder how we get half of these odd beasts!!

158) Revenge (1986) Obscure title about citizens in Tulsa, Oklahoma; who worship a legendary demon-hound.

159) Escapes (1986) A " Twilight Zone" type of series. "Who's There?" is about a jogger who meets elven creatures.

160) Body Count (1986) More gore. Explorers are picked off by a legendary Bigfoot. The uncut version is banned.

161) Serpent Warriors (1986) A construction site is plagued by a lost tribe, who worship a legendary serpent.

162) **The Night of the Sharks** (1987) Another turgid giant shark film.

163) **Leviathan** (1989) Metamorphosis movie. A crew of underwater miners investigate a Russian ship, and turn into serpents.

164) **Guinea Pig** (1990s) A series of very sick and depraved movies appeared under this title, one being called "Live Mermaid in a Man-hole." Extremely disturbing and unavailable.

165) **Missing Link** (1992) Michael Gambon narrates this docu-drama on primitive man and his survival.

166) **Bigfoot: the Unforgettable Encounter** (1994) Rather like "Bigfoot and the Hendersons"; an enchanting children's tale of a child's relationship with a hunted Sasquatch.

167) **Cry Wilderness** (1994) More Bigfoot soppiness as a child befriends a man-beast.

168) **Baboon** (1994) A low-budget tale about a Yeti.

169) **Last of the Oso Si Papu** (1994) A lost tribe of lizard-men in another low-budget film.

170) **They Bite** (1995) Awful home-movie type film involving aquatic killers, lots of nudity, jokes and stupid Americans.

171) **Winterbeast** (1995) A frustrated Bigfoot terrorises a ski-resort.

172) **Anaconda** (1996) New giant snake movie with a big budget.

173) **Little Bigfoot** (1996) Another children's film about a cute bigfoot. This film also spawned a sequel.

174) **The Flying Serpent** (no date) This film contains a different kind of beast; a flying fox with fangs.

175) **The Beginning of the End** (no date) A city is horrified by the existence of giant grasshoppers.

176) **The Phantom Empire** (no date) An expedition invades the forgotten privacy of dinosaurs.

177)   The Lake Worth Monster (no date) A stage play built around a famed lake monster.

*   *   *

Film chronicles are becoming more difficult to compile; but a new search begins as TV programmes now feature crypto-related creatures. The films usually use the beast as the topic; whereas in TV they only feature briefly in a series or soap. We all remember the "Chimera" drama , although it did not use the real legend. In "First Born", an ape boy was conceived. "The X Files" churned out "Big Blue", the rather uninspired tale of a lake monster. Even "Emmerdale" cashed in on the "Beast of Bodmin" case, by having its own "Beast of Beckinsale"; which turned out to be a boar. "Heartbeat" also had a mystery beast on the prowl; and despite the rumours of a panther, was revealed as a wild dog.

In the past, "The Munsters" and "The Goodies" paid homage to the Loch Ness Monster, while "Dynasty" had problems with Bigfoot. Some crypto-material has also found its way into TV advertising. In 1994, an advert for "Tab," a soft drink, included a sequence from the Patterson Bigfoot film. 1996 brought a certain car advert, with UFO footage, Nessie and Bigfoot. Both "Vodaphone" and "Cheltenham & Gloucester" used adverts containing Yetis in 1997. Bigfoot was mentioned in the horrendous "California Dreams" and was seen rummaging through the rubbish in "Eerie Indiana". The BBC TV series "Goosebumps" featured an "Abominable Snowman of Pasadena" episode. Meanwhile, Sony Playstation has a newly-released game called "The Jersey Devil". There appears to be an increasing quantity of crypto-material finding its way into children's

programmes, so perhaps this will be the location for our next quest.

If anyone can think of any further crypto-related programme or film then please write to me at :-

8, Gorse Avenue,   Weedswood Estate,
CHATHAM,   Kent,   ME5 0UQ.

I do realise that there are films which I have forgotten, but this is the list that challenges the world, and proves to be the ultimate collection. If anyone would like to view some of these films or obtain any crypto-documentaries or news clips, then please write to the address above. I would just like to thank "Creature Features" and John Stanley. I would also like to thank Vicki for her crypto-cartoon knowledge, and my forever angel Nicki. For now, happy reading and happy hunting.

Neil Arnold

Chupacabras - Nothing New

by Terry Hooper

My Texan colleague, Lindsay Whitehouse, has called me "very knowledgeable" on UFOs and their kith and kin. I tend to think that my hyper-active mind seeks out these oddities to stop me going mad in a mundane world. At times, however, that stored knowledge pays off; as in the case of the Chupacabras.

I have extensive files on UFO / non-UFO entity incidents around the world going back to 1900; and these came in handy when, as a member of Ivan T. Sanderson's international Society for the Investigation of the Unexplained (S.I.T.U.), I set up a small bureau in the U.K. (still going). Having seen brief items in "Fortean Times," "UFO Magazine," etc., about the Chupacabras; something in the back of my mind said that this all seemed to be very familiar. So I reached for the South and Central America files.

There it was. "There what was?" - It began in 1973, as far as I can tell. From 18.00 hours on the 20th October to 06.00 hours on the 21st October, at El Yunque Mountain in Puerto Rico, no less than nine people saw four "weird" creatures, each of about 5-6 feet (1.5-1.8 m) in height. These creatures were quite active on the mountain slopes and made strenuous efforts to avoid the light of torches shone at them. Branches of trees were found broken and also strange footprints. [1 & 2]

Nothing more happened after this, as far as I know, until 20.00 hours on the 18th April, 1975. At Ponce, Orlando Franceschi, a farmer, took a shovel out into the backyard of his home, after seeing what he took to be a stray dog. But what the farmer came face to face with was no dog. It had "long ears, a long nose to the mouth, which was slit with no lips; two black blobs for eyes and the jawbone of an ape." When this creature walked, it did so swaying from side to side. This strange animal was about 4ft 8ins (1.4 metres) tall. Franceschi felt that the creature was going to attack, so he struck it a hard blow in the chest and it backed away. He hit it again, but at the third swipe Franceschi fell as though he had missed his target and landed on the ground, paralysed ,though this may have been more through fear or shock. The creature had gone and the farmer soon regained his strength. During this incident his dog had not barked once.

After Franceschi's encounter, five young men saw a "funny little man" and pelted it with stones. A week later, Senor Franceschi imagined he heard voices, which came to nothing. [3]

The next known incident, took place at 20.30 hours, 12th July, 1977 at Quebradillas. Senor Adrian de Olmos Ordonez, was taking a rest on his balcony, when he suddenly saw something emerge from the grounds of a farm opposite his house. It was a 3ft 6ins (90cms) tall figure, that walked in normal fashion towards a street lamp. The man called his daughter, Irasema, to bring a pencil and to turn on the room light, so that he could sketch the being. Irasema accidentally turned on the balcony light instead, which seemed to frighten the creature, which was now close to the street light. The creature ran back under the farm's barbed wire fence and then stopped; placed its hands on the front part of its belt and the rucksack-like object on its back lit up. A high pitched noise followed and the creature flew up into the trees. By now the neighbours and family members of the witness had arrived and saw lights moving about in the trees.

The creature was said to be normally proportioned, except for the arms, which appeared to be rather short. No real idea of height was given. The entire body was covered in a garment that was green in colour and looked as though "full of air". The head was covered by a metallic helmet with a glass visor with a small lighted 'aerial' above it, and " there were things sticking out that seemed like pointed ears". The light reflecting off the visor made it impossible for the witness to see the facial features. A tail was also noted. Apart from the suit, one may have thought that the creature was a monkey. The sighting lasted for 60 seconds, so the possibility that the witness got some details wrong, due to either psychological factors or lighting conditions, must be considered. Certainly cattle and dogs seemed disturbed and police guards were later posted until daylight. [4]

On the 4th September 1977, a similar creature was seen by a 74 year old farmer at Barrio Abra Centro, near Corozal [5]. Although I can't find a more reputable English source, at Uyuni, Bolivia, in 1968, a Senora Valentina Flores found herself confronted by a similar creature after 34 of her sheep had been killed [6]

But the real clincher comes in 1976. In that year the "Flying Saucer Review" (vol 22 ; nos 5 & 6) [7] carried articles by Sebastian Robiou-Lamarche on "UFOs and Mysterious Deaths of Animals". Robiou-Lamarche puts the start of mysterious animal deaths on the island as 1975; almost two decades before the Chupacabras phenomena.

Between February and July 1975, there were a large number of animal deaths in the same geographic areas as sightings of UFOs; though this may be only circumstantial, depending on your viewpoint. The first deaths took place on 25th February, in the area around the town of Moca in the north-west of the island. By March there was already a sensationalist name for the killer, "the Vampire of Moca". As with the Chupacabras, calls were made for Government action. At the end of March, reports had started to come from Aguadilla and were spreading to other areas. Snakes were ruled out and the idea of a blood-drinking madman soon vanished also. In July, vampire bats were also ruled out by experts.

The capital, San Juan, experienced animal deaths in April, while Moca experienced more in July. No one could offer an explanation.

The facts were these; all the animals had been killed at night, and usually in the early morning hours. In almost every case, the owner of the animals heard nothing, even when sleeping close by. In those cases where the owners were disturbed, it was by a "loud screech" or a sound similar to that of wings beating. In some cases a strange animal was seen. This killer left wounds on its victims, varying in accordance with the size of the creature; small wounds on birds, large wounds on goats. No blood was found anywhere around the wound; which was left open, as though whatever caused it took away any flesh or organs, encountered in the process. Positions of wounds varied, but were mainly in the neck and thorax areas. Some animals had their necks completely broken. Killings were selective; in pens where there were other birds or animals, only one species was killed; the others had no signs of wounds or attack.

The list of animals killed and their percentages were:-

Domestic Fowl	182 (58.14%)	Cows	8 (2.56%)
Ducks	40 (12.78%)	Sheep	5 (1.60%)
Goats	33 (10.54%)	Pigs	3 (0.96%)
Rabbits	20 (6.39%)	Dogs	3 (0.96%)
Geese	18 (5.75%)	Cats	1 (0.32%)

That means that the bulk of the victims (83.07%) were those kept in pens and hutches. Killings were in rural and suburban areas, but what of the strange creatures seen?

Some witnesses reported "a strange animal, very hairy, running away--" or a "--screech, as though from a gigantic bird" was heard; other sounds included "a loud hum", "a deafening noise" and "a loud flapping noise". One has to be cautious over how sounds are described. Don Cecilio Hernandez (65) lost a total of 35 chickens over several nights and on the last of these occasions saw "what looked like a woolly dog-- with no legs or head-- running off towards the hills silently-- it looked just like a mass of wool running along." Obviously, if it was running, it had legs. There were many odd animal sightings not directly related to animal deaths.

In early March, at around 00.30 hours, Maria Acevedo, of the Barrio de Maria district, Moca, heard "a strange animal on the zinc roof of her house" and could hear it walking about and pecking before it flew off with a "terrible screech". On 25th of March, Pellin Marrero of Rexville, Bayamon, reported a "whitish-coloured gigantic condor or vulture" flying around. Juan Munis Feliciano, Barrio Pueblo, La Sierra sector, Moca, reported being attacked --"by a terrible greyish creature with lots of feathers, a long thick neck, bigger than a goose;" on the 26th of March at 22.00 hours. Neighbours heard his calls and drove the bird away with stones. The same day, Olga Iris Rivera and Barbara Pantoja, of Nemesio Canales housing complex, saw " a gigantic bird flying around among the clouds."

The whole affair was never explained. Other reports in 1975 also came from Miami, and other parts of the United States where, oddly enough, Chupacabras events are taking place today. Reports of creatures were not uncommon in the late 1970's; for example the "Dover Demon". There appears to be substantial evidence for a large species of bird of prey, effecting a quick strike, giving the victim little time for escape or effective resistance. Unfortunately, the majority of victims were in pens and hutches; and therefore inaccessible to this type of predator. I've yet to see a bird of prey land, walk over to a cage, open it and kill its victim. However, the style of killing is not consistent with a bird. [8 & 9]

A link between UFOs and the animal deaths is unlikely; there is not a shred of evidence. Quebradillas seems inconsistent with the other creature reports. It may be that, if investigators delve back into newspaper archives of the 1973-75 period; they will find either better descriptions of these or other "vampire killings". The Chupacabras is nothing new; but the question which needs to be investigated more deeply is whether it is a previously uncatalogued Puerto Rican animal, or something much more sinister.

References

1. Awareness, vol 3 no 2, 1974.
2. UFO Register, vol 5 nos 1/2, 1974.
3. FSR vol 22 no 6, 1976.
4. FSR vol 23 no 6, 1977.
5. FSR vol 24 no 2, 1978.
6. Official UFO Nov 1976.
7. FSR vol 22 nos 5 & 6, 1976.
8. Pursuit, vol 13 no 1 (whole no 49) Winter 1980.
9. Clark, J. and Coleman, L; Creatures of the Outer Edge, Warner Books, N.Y. 1978

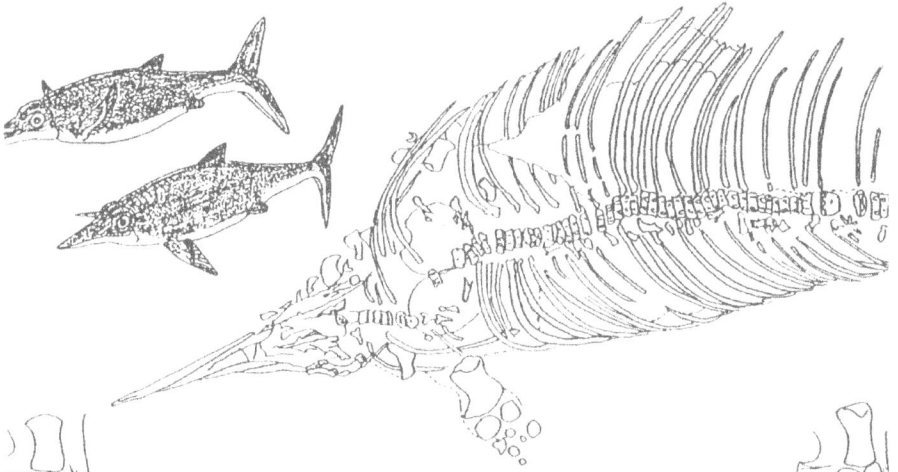

Aspects of Ichthyosaur Evolution and Ecology

With Comments on
Cross-taxon Convergence Seen Throughout Marine Tetrapods

by

Darren Naish

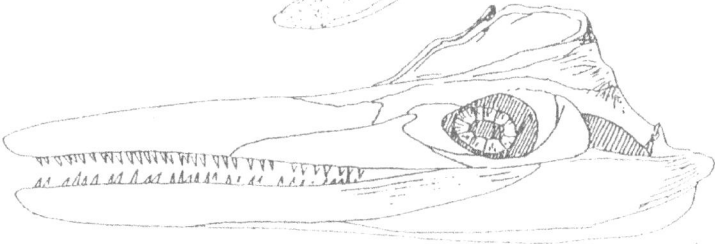

Aspects of Ichthyosaur Evolution and Ecology
With Comments onCross-taxon Convergence Seen Throughout Marine Tetrapods

INTRODUCTION: The Bringing Together of the Marine Tetrapods

Marine tetrapods - that is, all four-limbed vertebrates that have secondarily invaded the marine realm - have anexcellent fossil record, and there is correspondingly a vastbody of relevant literature. They represent successive invasions that have occurred repeatedly over the last 250 million years and include numerous unrelated groups that have often adapted to marine life in similar ways.

Because of the constraints on morphology and ecology imposed by marine life, marine tetrapods offer some of the best models of convergence and parallelism yet found in nature. What is little appreciated is how striking some of this cross-taxon convergence is. A paradigm of 'neglected analogues' is found to be the best approach where remarkably similar adaptations amongst unrelated groups are hypothesised to have evolved for the same function. This is essentially identical to the so-called Kowalevskian approach discussed recently by Bakker (1996), but was developed independently. Well known examples include the wing-like flippers of plesiosaurs, sea-lions and penguins, and the shark-like profile of ichthyosaurs and dolphins (fig. 1).

Analogies amongst functional morphology and by inference behaviour and ecology in the marine tetrapods are far more extensive than this. This area has generally not received much attention in the literature because hardly any authors have been able to extend their coverage beyond the boundaries of one or two groups. Those interested in marine reptiles have rarely been able to note the parallels provided by whales or seals while students of marine mammalogy have not typically had good reason to examine the parallel a Triassic reptile might provide for an extant mammal. An important recent exception is that provided by Taylor (1987), who noted that superficially similar cranial designs, e.g. the 'forceps jaw' morphology, have been evolved repeatedly by marine tetrapods in order that they overcome the problems of catching, handling and dismembering prey in water. The example of the 'forceps jaw' design appears in crocodiles, ichthyosaurs, cetaceans, and to a degree in some aquatic birds and plesiosaurs (fig 1). Examples such as this are widespread. They include the robust, big-toothed skull design seen in macropredatory marine tetrapods adapted for the seizing and tearing of large, bony prey in water - the sharp-toothed, pointed skull design exemplified by sharp toothed, piscivorous plesiosaurs and pinnipeds, and the broad, powerfully muscled skulls with modified palates seen in the specialised placodonts, odobenids, and one unusual cetacean. Occasional comments in the literature have been made on the subject of the similarity of these morphologies, but they have never been compared side by side. Collation of morphometric data recorded from all available taxa would provide the framework for a technical study of morphology-function correlation and pave the way for recognition of macroevolutionary patterns as recurrent themes seen throughout the long history of

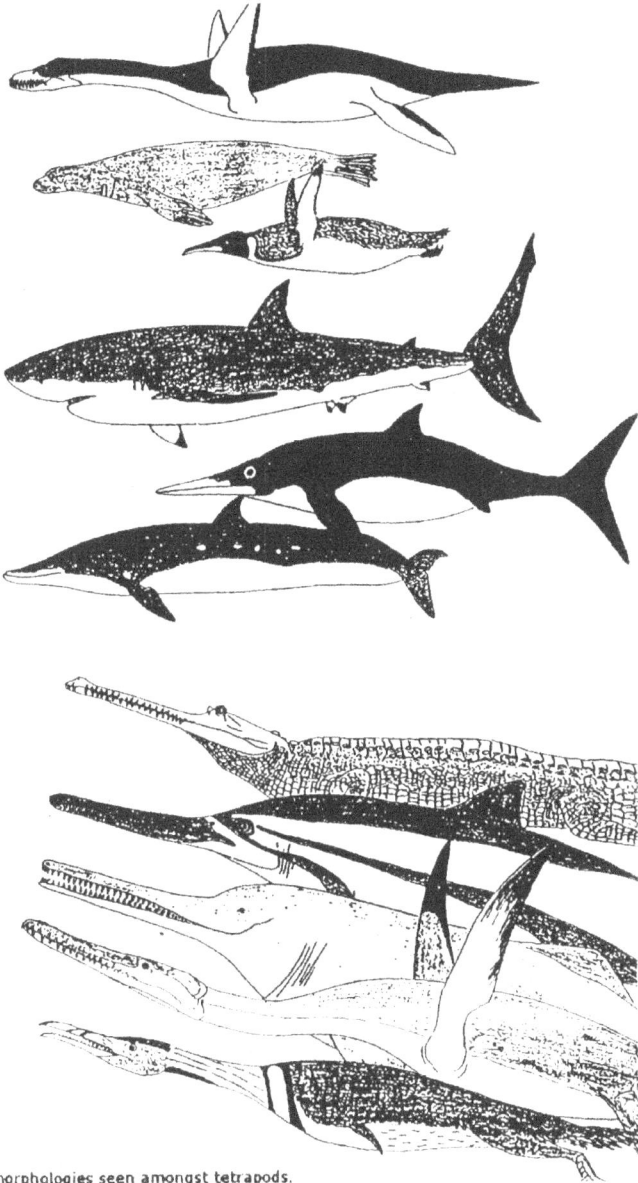

Fig. 1. Convergent morphologies seen amongst tetrapods.

(a) Convergent body plans; underwater fliers, exemplified by (top to bottom) a cryptoclidid plesiosaur, otarid and spheniscid; and thunniform swimmers, illustrated by a lamnid shark, ichthyosaur and delphinid cetacean.

(b) The 'forceps jaw' morphology as evolved convergently in (top to bottom) crocodiles, ichthyosaurs, cetaceans, plesiosaurs and birds.

tetrapod adaptation to life in water.

This contribution was initially designed as an elementary attempt directed towards such an end but the scope proved too large. The macroevolutionary slant will be commented upon in this work, but only as a series of observations and without detailed investigation of all applicable taxa. The history, functional morphology, and palaeoecology of but one marine tetrapod group, the Ichthyosauria, forms the basis for the present work.

Reconstructing Palaeocommunities.

Some other aspects of marine tetrapod evolution have not been dealt with adequately to date. As with the morphologies mentioned above, community structure is an area that could be analysed by collation of existing literature. Substantial reviews must be published before any single, coherent view of trends may emerge, and it is simple 'bringing togethers' such as this one that will represent the first step towards the contruction of chronofaunal models.

Only one significant study of community structure in a Mesozoic marine fauna has been published: that on the Oxford clay fauna of Peterborough by Martill *et al.* (1994). This example proved a striking parallel with modern marine tetrapod communities, as noted by McHenry and Naish (1996). Their approach, to find analogues amongst the extinct and extant taxa based on diet and comparative morphology, can easily be extended to other marine communities, once databases of taxa, dietery preferences and reconstructed ecologies and behaviours are compiled. By studying the ecology and functional morphology of extant forms, research made easy by consultation of good texts on modern marine mammals (Bonner 1994; Evans 1993; Jefferson, Letherwood and Webber 1993; King 1983; Ridgeway and Harrison 1981*a*, 1981*b*, 1985, 1989, 1994), enough can be guessed about Mesozoic marine tetrapods to restore their place in a community.

Some attention has been given to community structure in the marine tetrapod faunas of the Caenozoic; those of the late Miocene in particular (Barnes 1976; Barnes, Rasche and McLeod 1984; Mitchell 1966). Again these prove to be largely analogous with modern faunas but with individual roles filled by different, more archaic taxa. It is noticeable that, in contrast to modern faunas but with individual roles filled by different, more archaic, taxa. It is noticeable that in contrast to modern communities of large terrestrial vertebrates, we seem extraordinarily fortunate in still having intact marine faunas.

Environmental controls such as circum-continental currents have been hypothesised as responsible for the radiation of some marine tetrapod groups. Fordyce (1977, 1980, 1989), has made the most important contributions in this area. He has suggested that the creation of the Circum-Atlantic Current in the Late Oligocene was responsible for an evolutionary burgeoning amongst marine mammals. Such models could be extended to the Mesozoic fauna by simple analogy but as yet there is no apparent correlation between the burgeoning of Mesozoic clades and any palaeoceonographic event. This is most probably an area that needs further attention and it seems likely that such correlations may exist even though events comparable with the creation of the Circum-Antarctic Current did not occur in the Mesozoic.

It seems apparent that when compared with their terrestrial contemporaries (dinosaurs and other groups for the marine reptiles; mammals such as elephants, horses and carnivorans for the marine mammals), marine tetrapods are generally neglected. It should be noted that Carrolls's textbook (Carroll 1988), includes, thankfully, a moderately substantial overview of marine reptiles, and Savage and Long (1986), the standard reference work on fossil mammals, has a brief but extremely competent treatment of marine mammals. In popular works, however, it is typical to find only one or two pages devoted to marine tetrapods (Benton (1990) is a true exception): a woefully inadequate representation in view of the historic importance of the group and the fascinating aspects of their ecology and biology that are known. Interest in marine tetrapods appears to be growing, however. Improved recognition and understanding in the field of marine reptile palaentology will occur with the publication of a much awaited volume, *Ancient Marine Reptiles*. This is a symposium volume resulting from a seminar organised by the American Society of Vertebrate Palaentology. At the time of writing it has not been published, but it is due out early in 1997.

Constraints in the Evolution of Marine Tetrapods.

Occasionally it has been suggested that certain marine tetrapods are direct descendents of amphibious or aquatic ancestors (e.g. Romer 1974). However, the accepted concensus today is that all known marine tetrapods (excepting certain amphibian grade tetrapods) descend from terrestrial ancestors. The same pattern of increasing independence from land can therefore be seen in all marine tetrapod groups and numerous parallels can be observed. Basal numbers of marine groups are generally amphibious, they tend to breed on shore, are largely restricted to coastal marine environments, and are not markedly specialised for aquatic propulsion nor are they of large body size. Their anatomies can be viewed as a kind of compromise. Extant lutrine Carnivora, and perhaps the Marine Iguana (*Amblyrhynchus cristatus*) appear to present us with suitable analogues for these indeterminate taxa, otherwise known only as fossils, so they should therefore be important objects of stydy. Marine lutrines have received fairly extensive coverage in the literature (reference lists can be found in Chanin (1985), Harris (1968), Kenyon (1981) and Love (1990)). By comparison *Amblyrhynchus* is positively neglected and only one technical paper, that by Dawson, Bartholemew and Bennett (1977) deals with its adaptations to the marine environment.

Constraints imposed by the ancestral plan shape the style of propulsion adopted and effectively predict the style of movement the fully marine, large bodied descendents of basal amphibious taxa will evolve. Caroll (1985) discussed this area, but, by comparing marine reptiles with their fully terrestrial nearest relatives, concluded that the swimming styles evolved did not recall the functional morphology of ancestors. This is a reflection of the modification marine lineages have undergone but, even so, it may be difficult to imagine vast evolutionary gulfs existing between terrestrial forms and their amphibious progeny. Comparisons of the functional morphology and ecology of basal taxa from different secondarily marine groups verify the predictive model. Descriptions of primitive cetaceans in recent years, for example, have shown that these animals experimented with a variety of lifestyles and morphologies (Gingerich et. al. 1994; Novacek 1994; Thewisson et. al. 1996), and foreshadowed advanced cetacean swimming by moving in vertical undulations with large feet providing the main propulsive thrust (Thewisson, Hussain and Arif 1994). This type of swimming (fig 2) encouraged generation of a dorsoventrally generated force from the tail and the evolution of caudal flukes resulted soon afterward (Gingerich, Smith and Simons 1990). Limb dominated locomotion in basal sauropterygians has similarly been identified as antecedent to the subaqueous flight developed by plesiosaurs (Storra 1993; Sues 1987; Taylor 1989). Notably, such primitive stages of marine invasion are missing from some marine tetrapod groups. Ichthyosaurs especially are not represented by such intermediate forms, leaving the area open for speculation.

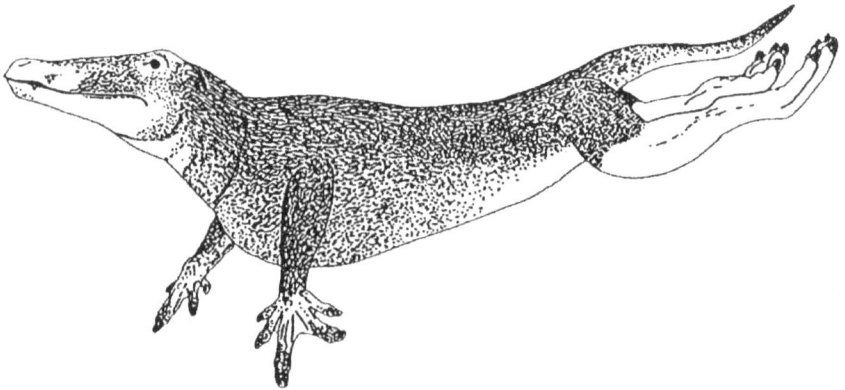

Fig. 2. Life restoration of the primitive lower-middle Eocene cetacean *Ambulocetus natans* (after Thewissen, Hussain and Arif 1994). Length 2.5 - 3 m. This animal swam by dorsoventral power strokes of the large feet and therefore foreshadowed the dorsoventral caudal-powered propulsion seen in advanced cetaceans.

Dependence on land in marine tetrapods is also constrained by other factors such as embryonic development, physiology and breeding behaviour. Pinnipeds, for example, may be tied to land because their moult phase and harem based breeding system render them physiologically and behaviourally unable to become truly independent of it. Some `logical` evolutionary outcomes may therefore be effectively impossible. An example may be a fully marine giant pinniped that gives birth at sea. Another example, a fully marine giant spheniscid (hypothesised by D.Dixon in Todd (1981) would not seem possible as avian embryos require time inside the egg for their lung membranes to dry. A somewhat similar situation is provided by development constraints on crocodiles and turtles whose embryos apparently derive their calcium from the egg shell and cannot therefore dispense with it (Tarsitino 1982). This also applies to birds (Burton 1991), and may be another reason why they cannot evolve live birth. In his discussion on the possible constraints on the evolution of marine mammals, Fordyce (1989) suggests that relative surface areas of structures or bodies may be limited by dictations of thermoregulation and/or buoyancy. As marine tetrapods are known to reach dimensions of 33 metres and 190 tons (Yochem and Leatherwood 1985), it is unclear what the upper size limit for a marine tetrapod might be, but, for endothermic marine tetrapods at least, a lower size limit is constrained by rate of heat loss. Downhower and Blumer (1988) calculated that 6.8 kg was the obtainable by such an animal - one less than or equal to the neonates of some extant small cetaceans.

Aspects of Ichthyosaur Evolution and Ecology.

The Geological Record and affinities of Ichthyosaurs.

Ichthyosaurs, the famous `fish lizards` of the Mesozoic are known from numerous complete, articulated skeletons that sometimes include soft tissue outlines. These make them amongst the most restorable of extinct tetrapods. The convergence of their bauplan with that of sharks, scombroids and cetaceans has been remarked on many times. Ichthyosaurs demonstrate a diverse spectrum of morphotypes, however, and whilst Jurassic and Cretaceous forms exhibit streamlined, thunniform bodies with forked or lunate tails (Martill 1996), the earliest, early Triassic forms, were elongate anguilliform swimmers (Motani, You and McGowan 1996). It should also be noted that the convergently similar thunniform groups were markedly different in some ways; McFarland et. al. (1979) note that sharks, cetaceans and inchthyosaurs all exhibit fundamentally different ways of exploiting their niche. A vast body of literature exists on ichthyosaurs and little of it has been reviewed. Excellent syntheses, however, are provided by Dechaseaux (1955), Carroll (1988), McGowan (1992 a), and Martill (1996). Howe, Sharpe and Torrens (1993) is a useful introduction to early discoveries and interpretations.

Ichthyosaurs of several diverse types are known from the late Scythian of the Early Triassic (Benton 1993; Massare and Calloway 1990; Mazin et. al. 1991), suggesting an earlier origin of the group. They thrived at high genetic diversity in the Late Triassic and Early Jurassic but did not retain diversity beyond the Toarcian (McGowan 1992a). In the Cretaceous they are predominantly represented by one genus, the wide-ranging *Platyptergius* (McGowan 1972a; 1972b). Its youngest known occurrence is from Upper Cenomanian rocks of Bavaria (Bardet, Wellnhofer and Herm; 1994) - Campanian and Hqannstrichian ichthyosaurs that have been reported (McGowan 1973a; 1978), are actually referrable to plesiosaurs (Baird 1984). Material apparently from an ichthyosaur is known from the Miocene of Malta (Ventura 1984), but there are doubts as to the specimen's true provenance. Indeterminate ichthyosaurian material from the Santonian was described by Teichert and Matheson (1944), and if it is a valid record, represent the youngest known ichthyosaur. It does not seem wise to extend the stratigraphic range of the group on the basis of some indeterminate bones, however, even though the material does not appear to have been reworked.

Why ichthyosaurs do not survive to the end of the Cretaceous is an oft remarked upon paradox (Halstead 1975; Stahl 1974), in view of the extreme modification to marine life of the group. It has been suggested that total ichthyosaur extinction ocurred at the Cenomanian/Turonian boundary as a result of a global ecological crisis (Bardet 1994). McGowan (1972a; 1973a), suggested that competition woth mososaurids was the responsible factor but at the time he thought that ichthyosaurids survived at reduced diversity into the Late Cretaceous. Mososaurids, do in fact, first appear in the Cenomanian (Benton 1993), apparently not long before the very last ichthyosaurs became extinct, but the markedly different ecological strategies of the two groups suggest that they would never have been in direct competition. Competition from large, predatory sauropterygians in the Late Jurassic and Cretaceous has been suggested as being at least partly responsible for a decline in ichthyosaur diversity. (Massare 1987).

The appearance of advanced neoselachians such as the lamniformes in the Early Cretaceous has aso been suggested as a possible explanation for ichthyosaurian downfall (Carroll 1988; Martill 1996). However, hardly any individual shark clade first appears at the right time to 'explain' ichthyosaurian extinction (Capetta, Duffin and Zydek 1993), and any possible correlation remains completely untested (as Hartill (1996) notes).

Ichthyosaurs appear to have been in decline regardless of shark evolution, and it seems highly improbable that simple interaction on the level of individuals can have major macroevolutionary effects (Benton 1983, 1988, 1991). Sharks have interacted with marine tetrapods for as long as the two have been contemporary but they have such an extensive and taxonomically complex fossil record that any discussion of them is well beyond the scope of this work. Cappetta (1987), is the best and most comprehensive review of Mesozoic and Cainozoic elasmobranchs to date.

Uncertainty continues to surround ichthyosaurian affinity and phylogeny. Dechaseaux (1955),

Appleby (1961), Gison (1971), and Massare and Calloway (1990) provide summaries of proposed ichthyosaurian affinities; these include nearly every group of reptiles, even including chelonians (Appleby 1959, 1961) and synapsids (Romer 1948). Huene wrote about ichthyosaur ancestry several times and correspondingly made a number of proposals as to affinity - his proposals include relatedness of ichthyosaurs to anthracosaur (cited in Olson 1971), and loxommatid (Huene 1944) amphibians, to mesosaurids (Huene 1922), and to microsaur-like forms (Huene 1952). Huene considered microsaurs to be reptiles at one time (Huene 1948): they are presently grouped with temnospondyls and are probably members of the same clade as the lissamphibians (Ahlberg and Milner 1994). Loxommatids are of unresolved position but share characters with anthracosaurs and both groups are probably successive outgroups to the amniotes (Ahlberg and Milner 1994; Lombard and Sumida 1992) as depicted in Figure 3. The most current view of mesosaurids is that they are the sister-group to Reptilia (Laurin and Reisz 1995).

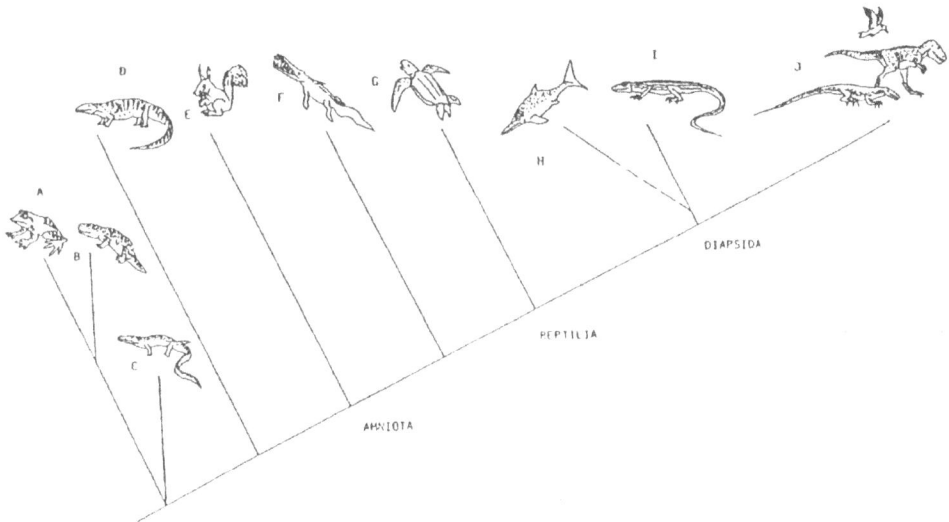

Fig. 3. A highly simplified cladogram of tetrapods that depicts relative positions of the groups mentioned in the text. They are (a) Temnospondyli + Lissamphibia, (b) Microsauria, (c) Loxommatidae, (d) Anthracosauria, (e) Synapsida, (f) Mesosauria, (g) Clelonia, (h) Ichthyosauria, (i) Younginiformes, and (j) Sauria. Ichthyosaurs are tentatively allied with younginiforms. Based on Ahlberg and Milner (1994) and Laurin and Reisz (1995).

The placement of all these taxa in modern cladistic frameworks emphasises the extremely tenuous nature of proposed relationships between these and ichthyosaurs - quite simply because ichthyosaur characteristics do not fit with those dictated by a non-reptile relationship. Some other proposed relationships such as those proposed by Neilsen's (1954, 1955), with trematosaurs, can similarly be rejected on the basis of character incongruity. Laenen's theory, that cetaceans are close relatives or descendants of ichthyosaurs (Sylvestre 1993), is absurd.

Riess (1986) argues that the extreme difficulty in recognising ichthyosaur ancestors, based on the presence of shared derived characters, may best be circumvented by deriving ichthyosaurs from "very plesiomorphic tetrapods". The same approach is taken in Riesss and Tarsitano (1989), where "early tetrapods" (namely seymouriamorphs) are suggested as the ichthyosaur sister-group.

Romer (1948) favoured ichthyosaur descent either from a hypothetical "pre-ophiacodont" or from a secondarily modified ophiacodont close to *Ophiacodon*. His argument was based on entirely primitive characters: the generalised amphibious morphology of *Ophiacodon* and the shared presence of a heavy stapes (Romer 1948). The latter is a characteristic of derived ichthyosaurs, however (Appleby 1959; Brinkman, Nicholls and Callaway 1992; Massare and Callaway 1990) and Romer's scenario does not withstand scrutiny.

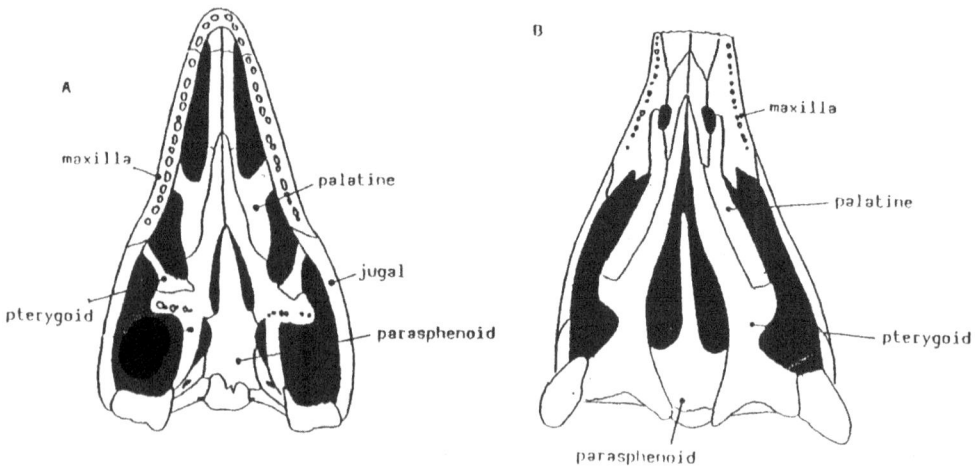

Ig. 4. Palatal configurations in (a) the younginiform *Youngina* and (b) *Ichthyosaurus* (both after Massare and llaway 1990). These are similar in a number of important ways (see text). Removal of the younginiform opterygoid would present an important step toward the ichthyosaurian condition.

As ichthyosaurs certainly have some of the captorhinomorph-diapsid listed by Lombard and Sumida (1992), an amphibian, synapsid or mesosaurid affinity can be excluded. Baur's suggestion (1887) that ichthyosaurs are diapsid reptiles has received increasing support in recent years (Carroll 1988; Massare and Callaway 1990; Nicholls and Brinkman 1995; Larsitano 1982) though exactly where ichthyosaurs would fit with regards to other diapsids remains mysterious. Massare and Callaway (1990) noted that ichthyosaur palates were similar in configuration to those of the younginiform *Youngina* (Fig.4) and that Triassic ichthyosaurs possessed all of the derived skull, palate and jaw characters of higher diapsids. These include a reduced lacrimal, posteriorly notched or emarginated quadrate, reduction or loss of pterygoid teeth, lack of parasphenoid teeth, loss of postspenial, and development of retroarterial process (Massare and Callaway 1990). The presence of these characteristics in ichthyosaurs further dismisses previously proposed relationships as discussed above. It was also found that ichthyosaurs shared more derived features with lepidosauromorphs than archosauromorphs (Massare and Callaway 1990) but at the time of this study Younginiformes were included amongst Lepidosauromorpha (Benton 1984, 1985; Evans 1984, 1988). Recent work indicates that younginiformes are outside of the neodiapsid crown-group Sauria (Laurin 1991). The same also may be true of ichthyosaurs. As the loss of the lower fork of the posterior jugal results in a boomerang shape (Fig.5) - the condition observed in both ichthyosaurs and lizards - it also argues for lepidosauromorph affinity, as may the loss in ichthyosaurs of the egg-laying habit (Tarsitano 1982). Because lizard (and it would seem ichthyosaur ancestor) embryos acquire their calcium rather from the yolk or the shell, they can dispense with shelled eggs (Tarsitano 1982). This argues indirectly for a lepidosauromorph affinity of ichthyosaurs, as we know without doubt that they gave birth to live young.

Carroll (1985) suggested, on the basis of a possible affinity with the archosaur-like hupehsuchians, that ichthyosaurs may be related to archosaurs. Hupehsuchians, themselves, are an enigmatic group that are not demonstrably related to archosaurs and there is presently no compelling evidence to show that they are related to ichthyosaurs either. McGowan (1992b) reported some unusual extraneural processes in two *Leptonectes* juveniles (Fig 6), that recall similar structures in *Hupehsuchus* as figured by Carroll and Dong (1991). There exists the possibility that these structures are homologous in which case they may support Carrol's (1985) suggestion of affinity. Hupehsuchians and Triassic ichthyosaurs do share a relatively long antorbital region, short transverse processes and a laterally processed thorax (McGowan 1992b) but whether these are convergently produced or shared derived characters is presently unknown.

Tarsitano (1982) thought that the ichthyosaur quadratojugal was anteriorly reduced and covered by a ventral extension of the squamosal He regarded previous extensions of this ventral extension as the quadratojugal (Andrews 1910; McGowan 1973b; Romer 1968) as an error: its identification as a ventrally extending squamosal is important as it also occurs in lizards with reduced lower temporal arcades. Massare and Callaway (1990) agreed with this approach and considered the

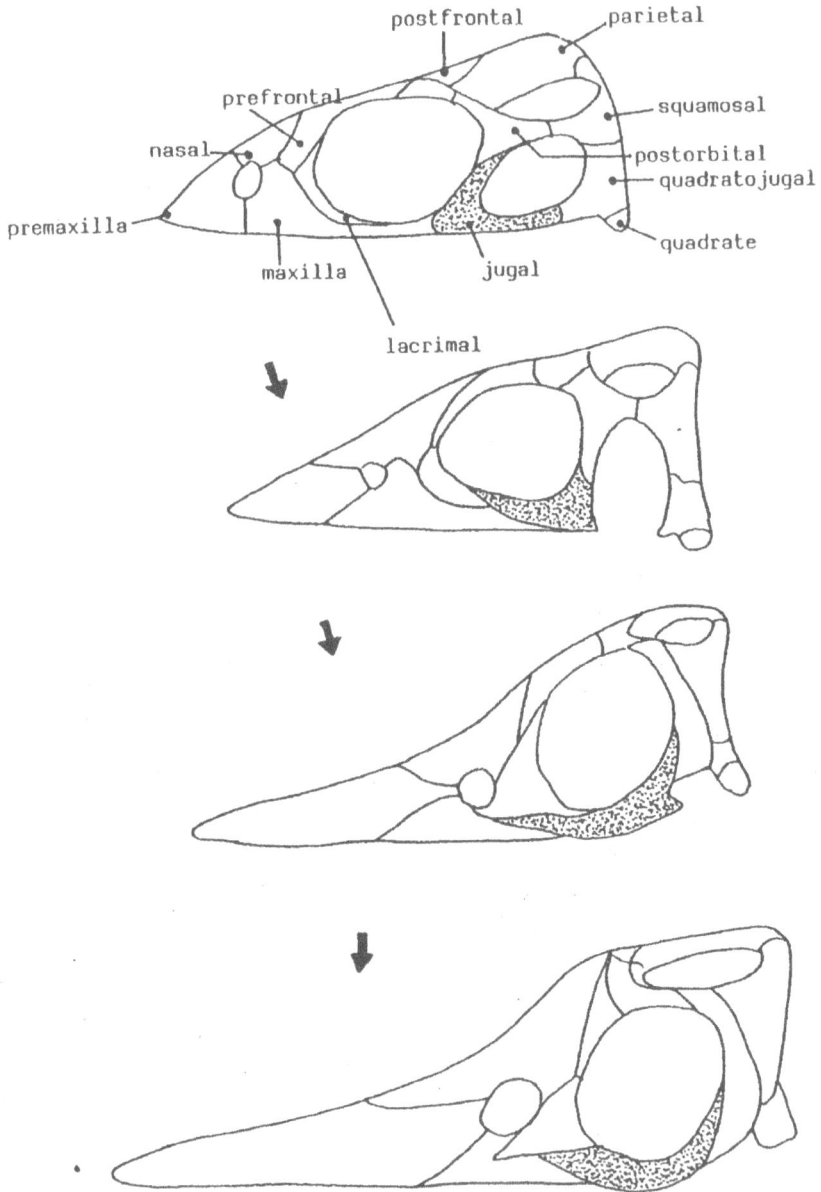

Fig. 5. Hypothetical stages in the evolution of the ichthyosaurian skull as reconstructed by Tarsitano (1982). The jugal (dotted) becomes boomerang-shaped as its contact with the quadratojugal is lost. See text for discussion. The identifications of some of the posterior skull bones given by Tarsitano (1982) are questioned by Nicholls and Brinkman (1995).

cranial region dorsal to the quadrate to consist of a combined squamosal and quadratojugal. However, Nicholls and Brinkman (1995) point out that both Sollas (1915) and McGowan (1973b) demonstrated that the quadratojugal could not be covered by the squamosal, and that the former was actually lost in all but the most primitive ichthyosaurs. A direct quadrate-squamosal contact with the quadrate foramen passing between the two is apparently retained from primitive diapsids where the quadratojugal is already reduced and restricted to the ventral margin of the skull (Nicholls and Brinkman 1995).

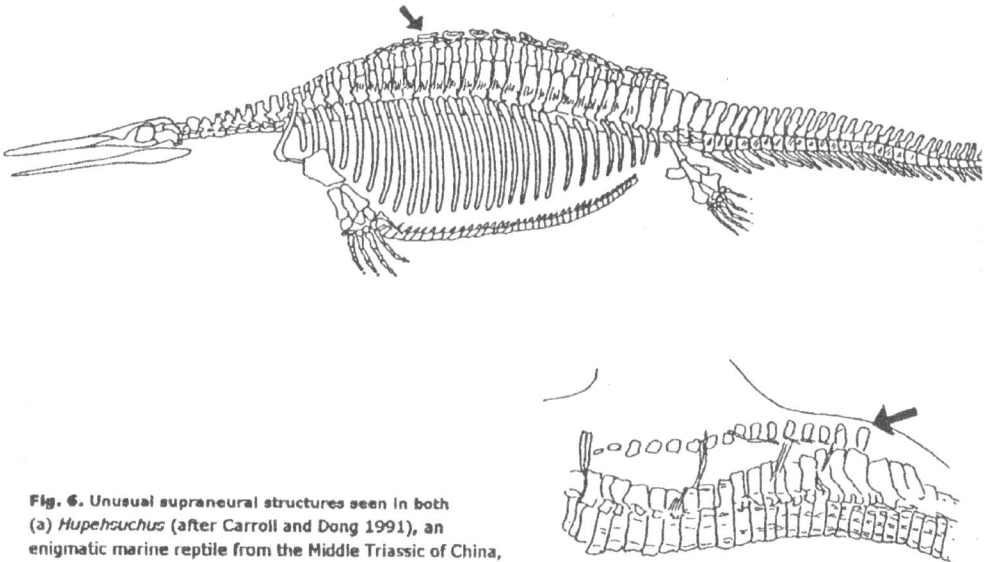

Fig. 6. Unusual supraneural structures seen in both (a) *Hupehsuchus* (after Carroll and Dong 1991), an enigmatic marine reptile from the Middle Triassic of China, and (b) juvenile ichthyosaurs (after McGowan 1992b). It is possible that these structures are homologous.

The conclusion adopted here, therefore, is that ichthyosaurs are close to younginiformes and are almost certainly neodiaspids (Fig. 3).

To mark both their unique morphology and the ambiguity inherent in previous phylogenetic theories, ichthyosaurs have traditionally been awarded their own reptilian subclass called Ichthyopterygia (e.g Camp, Allison and Nichols 1964; Carroll 1988; Dechaseaux 1955; Latarinov 1964; Wade 1990). Mazin (1981) has even raised rank of the Ichthyopterygia to class level and employed a restricted Ichtyosauria that is nested within Ichthyopterygia (Fig. 7).

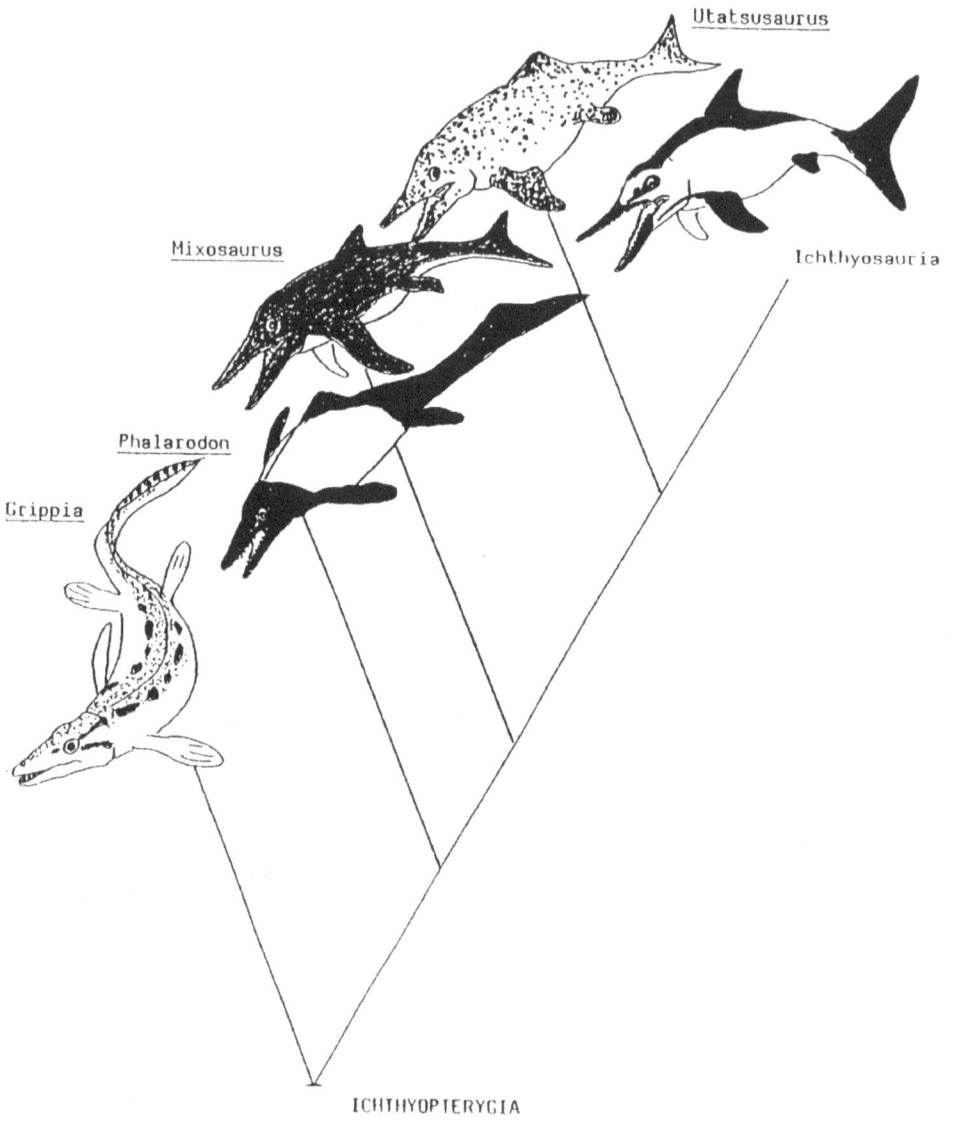

Fig. 7.

Whatever ichthyosaurs are it has also proved difficult to resolve relationships among them. Another tradition has been to divide them into two groups - latipinnates and longipinnates - an area best reviewed in McGowan (1973c). Recognition of the two groups began with V.Kiprijanoff in 1881 and was adopted and eventually applied to all ichthyosaurs with the publications of Lydekker (1891), Huene (1922, 1923), McGowan (1972c), and Appleby (1979). McGowan (1972c) was able to find both shared forefin and cranial proportions that distinguished the two groups. Longipinnates had three primary digits, three distal carpal elements, one digit supported by the intermedium, no digital bifurcation, widely spaced digital phalanges, and relatively large phalanges which were low in total number (McGowan 1972c). In latipinnates there were four primal digits and four distal carpal elements, two digits were supported by the intermedium, digital bifurcation was the norm, the distal phalanges were not widely spaced, and the phalanges were both small and numerous (McGowan 1972c). Figure 8 depicts relative latipinnate and longipinnate forefins.

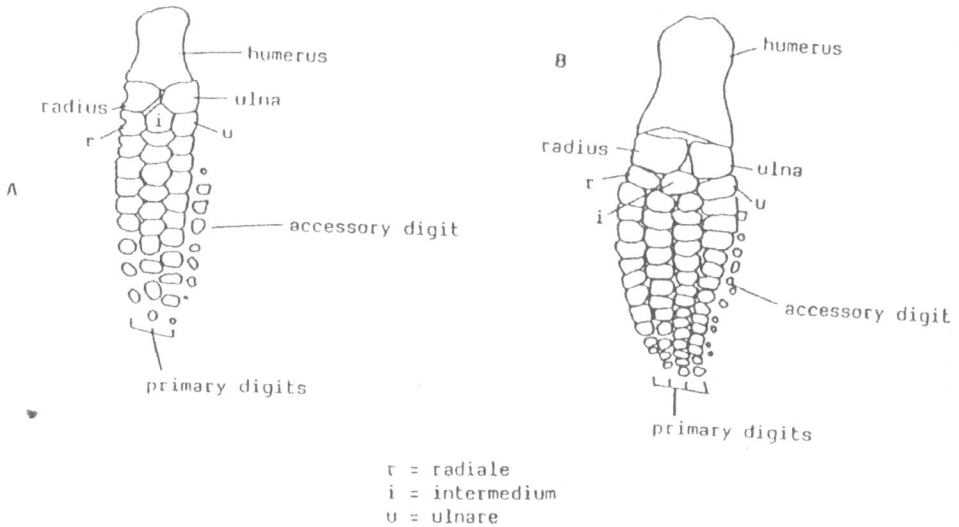

r = radiale
i = intermedium
u = ulnare

Fig. 8. Representative (a) longipinnate and (b) latipinnate forefins, after McGowan (1972c). See text for distinguishing characters of the two.

Appleby (1979) added complexity to the division by describing ichthyosaur forefins that did not correspond with either of McGowan's (1972c) definitions, and erected a diverse group called the Heteropinnatoidea. Members of this group presented a mosaic of intermediate features and showed that the latipinnate-longipinnate division was unreliable. However, doubts were not cast until McGowan (1976) assessed ichthyosaurian operational taxonomic units (OTU's), by performing multivariate alalyses on complete ichthyosaur skulls. Four primary clusters were produced. McGowan (1976) did not feel that the analysis was thorough enough in its coverage of taxa or morphology to rpovide a sound basis for a new taxonomy, but, by grouping together traditional latipinnates (e.g. *Mixosaurus*) with longipinnates (e.g. *Ichthyosaurus*) at a high similarity level, it did demonstrate that the latipinnate-longipinnate division was questionable.

Further suspicion was aroused by McGowan (1979) when latipinnate characters were found to exist in supposed longipinnates, and he eventually concluded, "it appears that there are no unequivocal distinctions between latipinnate and longipinnate ichthyosaurs, and that a systematic dichotomy of the group is probably unjustified." (McGowan 1979, p.126). The latipinnati and longipinnati have thus "died a natural death" (McGowan pers. comm. 1997).

REFERENCES

AHLBER, P.E and MILNER, A.R. 1994. The origin and diversification of early tetrapods. *Nature*, 368, 507-514.

ANDREWS, C.W. 1910. *A descriptive catalogue of the marine reptiles of the Oxford Clay*. Part 1. British Museum (Natural History), London.

APPLEBY, R.M. 1959. The origins of the Ichthyosaurs. *The New Scientist*, 6. 758-760.

APPLEBY, R.M. 1961. On the cranial morphology of ichthyosaurs. *Proceedings of the Zoological Society of London*. 137, 333-370.

APPLEBY, R.M. 1979. The affinities of Liassic and later ichthyosaurs. *Palaeontology*, 22 921-946.

BAIRD, D. 1984. No ichthyosaurs in the Upper Cretaceous of New Jersey or Saskatchewan. *Mososaur*, 2, 129-133.

BAKKER, R.T. 1996. Convergence between sabre-tooth mammals and Jurassic allosaurids - the Kowalevskian programme. *Journal of Vertebrate Palaeontology* 16, 21.

BARDET, N. 1994. Extinction events among Mesozoic marine reptiles. *Historical Biology*, 7. 313-324.

BARDET, N., WELLNHOFER, P and HERM, D. 1994. Discovery of ichthyosaur remains (Reptilia) in the Upper Cenomanian of Bavaria. *Mitteilungen der Bayerischem Staatssammlung für Palaontologie und historische Geologie*. 34 213-220.

BARNES, L.G. 1976. Outline of eastern North Pacific fossil-cetacean assemblages. *Systematic Zoology* 25, 321-345.

BARNES, L.G., RASCHKE, R.E and McLEOD, S.A. 1984. A late Miocene marine vertebrate assemblage from southern California. *National Geographic Society Research Reports*. 21 13-120.

BAUR, G. 1887. On the morphology of ichthyosaurs. *American Naturalist*. 21 837-840.

BENTON, M.J. 1983. Dinosaur success in the Triassic: a noncompetitive ecological model. *Quarterly Review of Biology*. 58 29-55.

BENTON, M.J. 1984. The relationships and early evolution of the Diaspids. *In* FERGUSON. M.W.J. (ed) *The structure, development and evolution of reptiles*. Zoological Society of London Symposium. 52. 572-596.

BENTON, M.J. 1985 Classification and phylogeny of the diapsid reptiles. *Zoological Journal of the Linnean Society*. 84. 97-164.

BENTON, M.J. 1988. The Late Triassic tetrapod extinction events. in PADIAN K (ed) *The Beginning of the Age of Dinosaurs*. Cambridge University Press, Cambridge, 303-320.

BENTON, M.J. 1990. *The Reign of the Reptiles* (Kingfisher Books, London).

BENTON, M.J. 1991. *Vertebrate Palaeontology*. (Harper Collins, London).

BENTON, M.J. Reptilia. *in* BENTON, M.J. (ed) *The Fossil Record 2*. (Chapman and Hall, London, 681-715).

BONNER, N. 1994. *Seals and Sea Lions of the World*. Blandford, London.

BRINKMAN, D.B., NICHOLLS, E.L. and CALLAWAY, J.M. 1992 New Material of the ichthyosaur *Mixosaurus nordenskioeldii* from the Triassic of British Columbia, and the interspecific relationships of *Mixosaurus*. *In* LIDGARD, S and CRANE P.R (Eds) *Fifth North American Plaentological Convention - Abstracts and Program. The Plaentological Society Special Publication*. 6, 37.

BURTON, R.W. 1991. Embryonic Development. *In* BROOKE M.deL. and BIRKHEAD, T.R (eds) *The Cambridge Encyclopaedia of Ornithology*. Cambridge University Press, Cambridge, 43-44.

CAMP, C.L., ALLISON, H.J., and NICHOLLS, R.H. 1964. Bibliography of Fossil Vertebrates. *Geological Society of America Memoirs*. 92, 1-647.

CAPPETTA, H. 1987. Mesozoic and Cainozoic Elasmobranchii. In SCHULTZE, H.P (ed) *Handbook of Palaeoichthyology, Chondrichthys II*, Volume 38. Gustav Fisher Verlag, Stuttgart and New York, 1-193.

CAPPETTA, H., DUFFIN, C., and ZIDEK, J. 1993. Chondrychthes. *In* BENTON, M.J. (ed) *The Fossil Record 2*. Chapman and Hall, London, 593-609.

CARROLL, R.L. 1985 Evolutionary constraints in aquatic diapsid reptiles. *Special Papers in Palaentology*. 33, 145-155.

CARROLL, R.L. 1988. *Vertebrate Palaentology and Evolution.* W.H.Freeman and Company, New York.

CARROLL, R.L. and DONG, Z. 1991. *Hupehsuchus,* an enigmatic reptile from the Jurassic of China, and the problem of establishing relationships. *Philosophical Transactions of the Royal Society of London.* B331, 131-153.

CHANIN, P. 1985. *The Natural History of Otters.* Christopher Helm, London.

DAWSON, W.R., BARTHOLEMEW, G.A. and BENNETT, A.F. 1977. A reappraisal of the aquatic specialisations of the Galapagos Marine Iguana *(Amblyrhynchus cristatus). Evolution* 31, 891897.

DECHASEAUX, C. 1955. Ichthyopterygia. *In* PIVETAU, J. (Ed) *Traite de Palaentologie, Tome V. Amphibiens, Oiseaux. Masson et Cie,* Paris 376-408.

DOWNHOWER, J.F. and BLUMER, L.S. Calculating quite how small a whale can be. *Nature* 335, 675.

EVANS, P.G.H. 1993. *The Natural History of Whales and Dolphins.* Academic Press, London.

EVANS, S.E. 1984. The classification of the Lepidosauria. *The Zoological Journal of the Linnean Society,* 82, 87-100.

EVANS, S.E. 1988. The early history and relationships of the Diapsida. In BENTON, M.J. (ed) *The Phylogeny and Classification of the Tetrapods, Volume One : Amphibians, Reptiles, Birds.* Clarendon Press, Oxford, 221-260.

FORDYCE, R.E. 1977. The Development of the Circum-Antarctic Current and the evolution of the Mysticeti (Mammalia:Cetacea) *Palaeogeography, Palaeoclimatology, Palaeoecology,* 31, 265-271.

FORDYCE, R.E. 1980. Whale evolution and Oligocene southern environments. *Palaeogeography, Palaeoclimatology, Palaeoecology,* 31, 319-336.

FORDYCE, R.E. 1989. Origin and evolution of Antarctic Marine Mammals. In CRAME, J.A, (ed) *Origins and Evolution of the Antarctic Biota.* Geological Society of London, London, 269-281.

GINGERICH, P.D., SMITH, B.H. and SIMONS E.L. 1990. Hind limbs of Eocene *Basilosaurus:* Evidence of feet in whales. *Science* 249, 154-157.

GINGERICH., RAZA, S.M., ARIF, M., ANWAR, M. and ZHOU., X. 1994. New whale from the Eocene of Pakistan and the origin of Cetacean swimming. *Nature,* 368, 844-847.

HALSTEAD, L.B. 1975. *The Evolution and Ecology of the Dinosaurs.* Eurobook Limited, London.

HARRIS, C.J. 1968. *Otters - A study of the recent Lutrinae.* Weidenfield and Nicolson, London.

HOWE, S.R., SHARPE, T. and TORRENS, H.S. 1993.*Ichthyosaurs: A History of fossil 'Sea Dragons'.* National Museum of Wales, Cardiff.

HUENE F von. 1922. *Die Ichthyosaurier des lias und ithre Zusammenhange.* Verlag Gebruder Borntraeger, Berlin.

HUENE F von. 1944. Die Zweiteilung des Reptislammes. Nues Jarbruch für Geologie und Palaentologie, Abhandlungen, 888, 427-440.

HUÉNE F von. 1948. The systematic position of the Microsauria. *American Journal of Science*, 246, 44-45.

HUENE F von. 1952. *Die Saurierwelt und ithre geschichtlichen Zusammenhange.* Gustav Fischer Verlag, Jena.

JEFFERSON, T.A., LEATHERWOOD, S. and WEBBER, M.A. 1993. *Marine Mammals of the World.* Food and Agriculture Organisation of the United Nations, Rome.

KENYON, K.W. 1981 Sea otter *Enhydra lutris. In* RIDGEWAY, S.H. and HARRISON, R.J. (eds) *Handbook of Marine Mammals Volume 1: The Walrus, Sea Lions, Fur Seals, and Sea Otter.* Academic Press, London, 209-223.

KING, J.E. *Seals of the World.* British Museum (Natural History), London and Oxford University Press, Oxford.

LAURIN, M. 1991. The Osteology of a Lower Permian eosuchian from Texas and a review of dipsid phylogeny. *Zoological Journal of the Linnean Society,* 101, 59-95.

LAURIN, M and RIESZ, R.R. 1995. A reevaluation of early amniote phylogeny. *Zoological Journal of the Linnean Society,* 113, 165-223.

LOMBARD, R.E and SUMIDA, S.S. 1992. Recent Progress in understanding early tetrapods. *American Zoologist,* 32, 609-622.

LOVE, J.A. 1990. *Sea Otters,* Whittet Books, London.

LYDEKKER, R. 1889. *Catalogue of Fossil Reptilia and Amphibia in the British Museum (Natural History). Part II. Containing the Orders Ichthyopterygia and Sauropterygia.* British Museum (Natural History) London.

MARTILL, D.M. 1996. *Fossils Explained,* 17: ichthyosaurs. Geology Today 12 (5), 194-6.

MARTILL, D.M., TAYLOR, M.A., DUFF., K.L., RIDING., J.B., and BOWN., P.R. 1994. The trophic structure of the biota of the Peterborough Member, Oxford Clay Formation (Jurassic), UK. *Journal of the Geological Society,* 151, 173-194.

MASSARE, J.A. 1987. Tooth morphology and prey preference of Mesozoic Marine reptiles. *Journal of Vertebrate Palaentology.* 7, 121-137.

MASSARE, J.A. and CALLAWAY, J.M. 1990. The affinities and ecology of Triassic ichthyosaurs. *Geological Society of America Bulletin.* 102, 409-416.

MAZIN, J.M. 1981. *Grippia longirostris.* Wiman 129, un Ichthyopterygia primitif du Tria inferieur du Sptisberg. *Bulletin de Museum national d'histoire naturelle.* 4. 317-340.

MAZIN, J.M., SUTEETHORN., V., BUFFETAUT., E., JAEGER. J.J. and HELMCKE-INGAVAT, R. 1991. Preliminary description of *Thaisaurus chonglakmanii* n.g., n.sp. new ichthyopterygian (Reptilia) from the Early Triassic of Thailand. *Comptes Rendus de l'Academie des Sciences, Paris. Serie II.* 313 1207-1212.

McFARLAND, W.N., POUGH, F.H., CADE, T.J., and HEISLER, J.B. 1979. *Vertebrate Life*, Collier Macmillan, London.

McGOWAN, C. 1972*a*. Evolutionary trends in longipinnate ichthyosaurs with particular reference to the skull and fore fin. *Life Sciences Contributions, Royal Ontario Museum*, 83. 1-38.

McGOWAN C. 1972*b*. The systematics of Cretaceous ichthyosaurs with particular reference to the material from North America. *Contributions to Geology*, 11. 9-29.

McGOWAN C. 1972*c*. The distinction between latipinnate and longipinnate ichthyosaurs. *Life Sciences Occasional Papers, Royal Ontario Museum*, 20, 1-8.

McGOWAN, C. 1973*a*. A note on the most recent ichthyosaur known: An isolated coracoid from the Upper Campanian of Saskatchewan (Reptilia:Ichthyosauria). *Canadian Journal of Earth Sciences*, 10. 1346-1349.

McGOWAN, C. 1973*b*. The cranial morphology of the Lower Liassic latipinnate ichthyosaurs of England. *Bulletin of the British Museum (Natural History), Geology*, 24. 1-127.

McGOWAN, C. 1976. The description and phenetic relationships of a new ichthyosaur genus from the Upper Jurassic of England. *Canadian Journal of Earth Sciences*. 13, 668-683.

McGOWAN, C. 1978. An isolated ichthyosaur coracoid from the Maastrichian of New Jersey. *Canadian Journal of Earth Sciences*. 15, 169-171.

McGOWAN, C. A revision of the Lower Jurassic ichthyosaurs of Germany with descriptions of two new species. *Palaeontographica Abteilung*, A166, 93-135.

McGOWAN, C. 1992*a*. *Dinosaurs, Spitfires and Sea-Dragons*. Harvard University Press, Cambridge, Mass.

McGOWAN, C. 1992*b*. Unusual extensions of the neural spines in two marine ichthyosaurs from the Lower Jurassic of Holzmaden. *Canadian Journal of Earth Sciences*, 29. 380-383.

McHENRY, C. and NAISH, D.W. 1996. A preliminary reconstruction of an Upper Triassic marine tetrapod community from Northwest Europe. In MAZIN, J. (ed) *Secondary Adaptation to Life in Water*. University of Potiers, Potiers. 24.

MITCHELL, E. 1966. Faunal succession of extinct North Pacific Marine Mammals. *Norsk Hvalfangst-Tidens*. 55 (3), 47-60.

MOTANI.R, YOU.H. and McGOWAN, C. 1996. Eel-like swimming in the earliest ichthyosaurs. *Nature*, 382, 347-348.

NICHOLLS, E.L. and BRINKMAN, D.B. 1995. A new ichthyosaur from the Triassic Sulphur Mountain formation of British Columbia. *In* SARJEANT, W.A.S. (ed) *Vertebrate Fossils and the Evolution of Scientific Concepts - A Tribute to Beverley Halstead*. Gordon and Breach Publishers, London, 521-535.

NEILSEN, E. 1954. Tupilakosaurus heilmani n.g., n.sp., An interesting Batrachomorph from the Triassic of East

Greenland. *Meddeleiser om Gonland.* 72, 133.

NEILSEN, E. 1955. *Tupilakosaurus.* In PIVETAU, J. (ed) *Traite de Palaentologie, Tome V: Amphibiens, Reptiles, Oiseaux.* Masson et Cie, Paris. 224-226.

NOVAKAK. M.J. 1994. Whales leave the Beach. *Nature,* 368. 807.

OLSON, E.C.1971. *Vertebrate Palaeozoology.* Wiley-Interscience, New York.

RIDGEWAY, S.II. and HARRISON, R.J. (eds). 1981a. *Handbook of Marine Mammals, Volume One: The Walrus, Sea Lion, Fur Seals and Sea Otter.* Academic Press, London.

RIDGEWAY, S.H. and HARRISON, R.J. (eds). 1981b. *Handbook of Marine Mammals, Volume Two: Seals.* Academic Press, London.

RIDGEWAY, S.H. and HARRISON, R.J. (eds). 1985. *Handbook of Marine Mammals, Volume Three: The Sirenians and Baleen Whales.* Academic Press, London.

RIDGEWAY, S.II. and HARRISON, R.J. (eds). 1989. *Handbook of Marine Mammals, Volume Four: River Dolphins and the Larger Toothed Whales.* Academic Press, London.

RIDGEWAY, S.H. and HARRISON, R.J. (eds). 1994. *Handbook of Marine Mammals, Volume Five: The first book of Dolphins.* Academic Press, London.

RIESS, J. 1986. Fortbewegungeweiss, Schwimmbiophysik und Phylogenie der Ichthyosaurier. *Palaeontographica.* A192, 93-155.

RIESS, J. and TARSITANO, S.F. 1989. Locomotion and Phylogeny of the Ichthyosaurs. *American Zoologist,* 29, 184.

ROMER, A.S. 1948. Ichthyosaur ancestors. *American Journal of Science,* 246, 109-121.

ROMER, A.S. 1968. An ichthyosaur skull from the Cretaceous of Wyoming. *Wyoming University Contributions to Geology.* 7, 27-41.

SAVAGE, R.J.G. and LONG, M.R. 1986. *Mammal Evolution an Illustrated Guide.* Facts on File Publications, New York.

SOLLAS, W.J. 1916. The skull of *ichthyosaurus* studied in serial sections. *Philosophical Transactions of the Royal Society of London.* B208, 63-126.

STAHL, B.J. 1974. *Vertebrate History: Problems in Evolution.* McGraw-Hill Book Company, New York.
STORRS, G.W. 1993. Function and Phylogeny in Sauropterygian (Diaspids) evolution. *American Journal of Science.* 293-A, 63-90.

SUES, H-D. 1987. Postcranial skeleton of *Pistosaurus* and interrelationships of the Sauropterygia (Diaspids).

Zoological Journal of the Linnean Society. 90, 109-131.

SYLVESTRE J-P. 1993. *Dolphins and Porpoises - A Worldwide Guide.* Sterling, New York.

TARSITANO, S. 1982. A model for the evolution of ichthyosaurs. *Nues Jahbruch fur Geologie und Palaentologie, Monatshefte,* 164. 143-145.

TATARINOV, L.P. 1964. Ichthyopterygia. *In* ORLOV, Y.A. (ed) *Fundamentals of Palaentology, Volume 12.* Gosudarstvennoe Nauchotekhnicheskoe Izdatel'Stfo Literaturi Po Geologii I Okhrane Nedr, Moscow 338-353.

TAYLOR, M.A. 1987. How Tetrapods feed in Water: A Functional Analysis by Paradigm. *Zoological Journal of the Linnean Society.* 91. 171-195.

TAYLOR, M.A. 1987. Sea-saurians for Sceptics. *Nature.* 338. 625-626.

TEICHART, C. and MATHESON, R.S. 1944. Upper Cretaceous ichthyosaurian and plesiosaurian remains from Western Australia. *Australian Journal of Science.* 6, 167-187.

THEWISSEN, J.G.M., HUSSAIN, S.T. and ARIF, M. 1994. Fossil Evidence for the Origin of Aquatic Locomotion in archaeocete whales. *Science,* 263, 210-212.

THEWISSEN, J.G.M., ROE, L.G. O'NEIL, J.R., HUSSAIN, S.T., SANHI, I., and BAJPAI, S. 1996. Evolution of Cetacean Osmoregulation. *Nature,* 381, 379-380.

TODD, P. 1981. After Man. *Wildlife,* 23 (10), 16-18.

VENTURA, C.S. 1984. The fosil herpetofauna of the Maltese Islands: A Review. *Naturalista Siciliana,* 8, 93-106.

WADE, M. 1990. A review of the Australian longipinnate ichthyosaur Platypterygius, (Ichthyosauria, Ichthyopterygia). *Memoirs of the Queensland Museum.* 28, 115-137.

YOCHEM, P.K. and LEATHERWOOD, S. 1985. Blue Whale *Balaenoptera musculus* (Linnaeus 1758). *In* RIDGEWAY, S.H. and HARRISON, R. (eds) *Handbook of Marine Mammals, Volume Three: The Sirenians and Baleen Whales.* Academic Press, London, 193-240.

An alphabetical listing of lochs, lakes, marshes and rivers associated with monsters.

by Michael Playfair

ABISDEALY LOUGH (LOCH AN BEISALAIG) (POUL NA GURRUM) - IRELAND, CO. GALWAY.

Length: 1 mile.   Breadth: 0.25 miles. This lough's Irish name apparently means 'lake of the monster' and has reports dating back to the Crimea war (1853-56); although I can find no more recent reports than 1914, the small number of reports seem to indicate a snake or eel-like creature. [1,2,3]

ACHANALT LOCH, SCOTLAND.

Length: 1050 ft (subject to periodic drainage which reveals large portions of its bottom). The supposed monster of this loch has apparently only one witness, a R. L. Cassie, who, in 1934, began observing a monster between 10 and 900 ft long! - and also six others between 100 and 200 ft. He had many sightings in the loch and also reported sightings in Lochs Cronn, Garve and Rosque. Either the whole episode was a hoax or Cassie suffered some form of mental disorder. [4]

AFANC LLYN YR (On R. Conway) GWYNEDD, WALES.

A creature called the 'Afanc' (sometimes taken to mean a beaver or a dwarf) living in the lake was apparently pulled out many years ago by a team of oxon, and dragged across to the Higher Lake Glaslyn - so that it could no longer cause floods. [5]

ALEXANDRIA LAKE, SOUTH AUSTRALIA

A creature called the "Moalgewanke" was described in 1879 by the Rev. G. Taplin. It was supposed to live in the lake and was said to resemble a mermaid with red hair, and had a voice like the rumble of a distant cannon. There are also reports from the surrounding terrain of giant monitor-like lizards. [1,6,7]

ALKAI LAKE, Nebraska, USA

I can only find one report from this lake, dating from either 1922 or 1923: Three men came upon an animal three-quarters out of the water near the shore, with an estimated length of 40 ft. Looking not unlike an alligator but with a stubbier head, it then moved further out into the lake. Cryptozoologist Roy P Mackal is of the opinion that it was a sea elephant. [1,8,9]

ALOA LAKE, MADAGASCAR.

The natives say that mermaids with hair reaching to their waists, called the 'Kalanoro', live at the bottom of this lake. [6]

AMALA RIVER, AFRICA

Home of a creature called the 'Oi-Umaina', described as 15 ft long with a head like a dog; small ears; short legs; claws; and a short neck. It is said to lie in the sun on the sand, and, when disturbed, slip into the water - leaving only its head visible. [6]

AMAZON RIVER, S. AMERICA.

Length: 4000 miles; drainage basin area: 2,250,000 sq. mi.

This, the second-longest river in the world, has reports of giant snakes - presumably **anacondas** up to 80 ft in length. [10]

ARKAIG LOCH, INVERNESS, SCOTLAND [adjoining Loch Lochy - q.v.]

Length: 12 miles. Mean breadth 1.5 miles. Max. depth: 420 ft. A few sightings from the 19th and early part of this century of a creature with a horse-head like head have been reported.

ASHBURTON RIVER, NEW ZEALAND.

Tracks said to have been discovered here in 1861 of an unknown otter-like creature called 'The Wattoreke'. [6]

ASSYNT LOCH, SUTHERLAND, SCOTLAND.

Famous for a article (said to have been written by Lord Ellesmere in the 1850s, which as far as I am aware has still not been located) about a monster which he said lived there. They may however be some truth in the matter as there have been reports from Loch Canish, which flows into Loch Assynt. [1,3,12]

ASTBURY MERE WATER PARK (Congleton), CHESHIRE, ENGLAND.

The mere is a flooded gravel pit approx. 15 years old with no streams or rivers flowing into it. In 1995 sightings of a 2.5 to 3 ft long lizard-like creature were reported - probably an escaped or released caiman or monitor lizard. Attempts at capture were unsuccessful and there have been no sightings since. [13,14]

ATTARIAF LOUGH, CORK, IRELAND.

Supposed sighting of a 10 ft long dark brown creature with a head like a calf was reported in the 1960s. [3,15,16]

AUNA LOUGH, GALWAY, IRELAND.

A glacial one-mile-long lake of unconfirmed depth, it has reports dating back to the 1900s of an eel-like creature up to 40 ft long, with humps and a mane. [2,3,15,16]

AUYAN-TEPUI (River on the Summit), VENEZUELA

This river on Table Mountain (said to have been the inspiration for Conan Doyle's "Lost World") seems to have produced a couple of reports by qualified observers - 3 ft long, long-necked creatures which have been called Pygmy Plesiosaurs. [7]

AWE LOCH, STRATHCLYDE, SCOTLAND.

This Y-shaped loch is one of the largest in Scotland. In the 18th century local people claimed that giant eels as big as a horse existed there. They were particularly noticeable in winter when it was said they could be heard crashing through the ice. [1,17,18]

AYLMER LAKE, QUEBEC, CANADA.

Connected to St Francis Lake, this lake was badly polluted in the 1950s, which led to a reduction in the fish population. There have been reports of an unknown creature - possibly a maskinonge (a type of North American pike). [19]

BAHR-EL-ARAB, AFRICA

There were reports earlier this century of a creature there called the Lau. [6]

BALA LAKE (LLYNTEGID), GWYNEDD, WALES.

At 4 x 1 miles, the largest natural lake in Wales. Max depth: 150 ft. Supposed home to a monster locally known as 'Teggy' or 'Anghenfil'.

Reports describe an 8 to 10 ft humped creature, or crocodile-like creature, or a more long-necked plesiosaur type. Janet and Colin Bord have suggested an extra-large fish as an identity and Dr Karl Shuker has reported that there were rumours that seals were put in the lake during the first world war for training for submarine detection. [5,13,20,21,22,23,24,25]

BALLYNANINCH LAKE, GALWAY, IRELAND.

Length: 2.25 miles. Apparently in the 1880s a lake monster was reportedly trapped here, under a bridge for a couple of days, because of a drought. It was described as a 30 ft giant eel as thick as a horse. Luckily for the creature, a sudden flood freed it - before the locals, who had planned to kill it, could do so. [3,15,16]

BANGWEULU LAKE, NORTHERN ZAMBIA, AFRICA

60 x 25 miles; area: 3800 sq mi. Two monsters reported. The first is described as smaller than a hippopotamus and Bernard Heuvelmans has speculated it may be a partially aquatic sabre-toothed

cal. The second is more reptilian in appearance, described as more of a sauropod type of creature. The name 'Chipekwe' has been used but there is some confusion as to which monster the name is applied. [1,6,7,26]

BAROMBI MBO LAKE, CAMEROON, AFRICA.

This large deep crater lake has a single report of two sauropod-like creatures, with necks 12 - 15 ft long and each with a capped horn-like projection on their heads. The report could be regarded as unreliable as it was not reported until nearly 30 years after the event, the main witness being, at the time, only 4 or 5 years old. [7,27]

BARWON RIVER, VICTORIA, AUSTRALIA

Reports in the 1850s said that Aboriginal natives believed that 'Bunyips' existed in the river, which they held in great dread. [28,29]

BATHURST LAKE, NEW SOUTH WALES, AUSTRALIA

5 x 4 km, very shallow - dries up in a drought.. A number of reports of creatures described as bunyips in the 1820s, usually described as about 5 ft long with a bulldog-like head, made a noise similar to a porpoise, and sometimes black flaps were re-observed to hang down around their necks. These sightings were probably of seals, as seals were discovered in 1947 in the Mulware River, separated from the lake by only 700 m of boggy ground. [6,29]

BATTLE RIVER, ALBERTA, CANADA.

In 1934 four people on different occasions were said to have seen a serpent-like monster approx 30 ft long. One witness apparently shot at it but the bullets bounced off it! [30]

BEAR LAKE, UTAH AND IDAHO, USA

Area: 109 sq mi. A number of reports in the 1860s and 1870s, sometimes described as up to 90 ft long and travelling at 60 mph! Some sightings were of a seal-like creature. It was also rumoured that a young monster had been captured in 1876, although there had been no further information and the report must be treated as suspect. [1]

BEISTLE LOCH, HIGHLANDS, SCOTLAND.

In the 1850s the landlord of this loch attempted to capture the monster said to reside there - first by poisoning the water and then by draining the loch. When this failed, it was reported he fined his tenants £1 each as punishment. [1,18]

LA BELLE LAKE, USA

Immense fish are said to have been seen in this lake. [31]

BENI, BOLIVIA.

An unknown creature was said to have been killed and apparently preserved in 1883. It was described as 12 m long with a flattened tail, and, unbelievably, 4 m behind the head, two more! Four short legs armed with claws and covered with scales. What happened to it and what one is to make of the creature is anyone's guess. [7]

BRAN LOUGH (BRIN LOUGH), KERRY, IRELAND.

1 x 0.5 miles. Max depth: 210 ft. This lough, said to be bare of fish, has a number of sightings of a monster. Some reports are even of the creature lying on the shore. Another report is of a creature 10 ft long looking like a cross between a dragon and a seal. [1,3,12,16]

BRAY LOUGH (LOWER), KICKLOW, IRELAND.

A sighting in 1963 by an anonymous witness of a large humped creature with a head like a tortoise, but much larger, has been reported. The report must be treated as suspect because of the anonymous witness, and perhaps to a lesser extent, the lack of any further sightings. [1,3,12]

BROCHU LAKE (on Govin Reservoir), CANADA

One report from here of a creature moving from dry land into the water. The witness attempted to catch it in his boat but was unsuccessful although he was travelling at 35 mph! He likened it to a prehistoric creature. [19]

BROMPTON LAKE, QUEBEC, CANADA

A number of reports were received from this lake in the 1970s. The creature was usually between 6 and 8 ft long with a head like a horse, with a moustache and possessing what seemed like ears. It also had a tail like a fish, and undulated. [19,32]

BROSNO LAKE, Twer (Ben Yok), RUSSIA

A number of sightings of the local lake monster with the name of 'Brosnie' (surprisingly...) were received in 1996. She was reported as 5 m long with an elongated neck. [33,34]

BULLARE LAKE, Sweden

Supposed monster said to exist here, with the body the size of a young calf and a neck 12 ft long. [1]

BURRUM (near Childers), Queensland, Australia

Reports date back to the beginning of the century of a long necked lake monster. [22]

BURREMBERT LAKE, VICTORIA, AUSTRALIA

Bunyip reported here in the 1870s.

CADER LLYN-Y, GWYNEDD, WALES

A report from the 18th century of a man who, whilst swimming in the lake, was pursued by a long object which raised its head and coiled itself around him, dragging him down to the bottom of the lake. [5]

CAMPBELL LAKE, SOUTH DAKOTA, USA

In 1934 it was said that a farmer was forced to ditch his tractor when a reptile-like creature crossed his path. Tracks were said to have been found, which disappeared into the lake. [9]

CANISP LOCH (FEITH AN LEOTHAID), SUNDERLAND, SCOTLAND.

This loch, which lies high above Loch Assynt (and to which it's a feeder) has a couple of undated sightings of a long-necked 'Nessie'-like creature. [12]

CAULOSHIELS LOCH, The Borders, Scotland.

Walter Scott reported an attempt to capture the supposed monster in this loch towards the end of the 18th century, and there is a report from the early part of the 19th of a creature which was like a 'horse or a cow' in the loch. [35]

CAW-LYND LAKE, WALES

Reported home of a water bull, described as having fiery horns, hooves, and flames issuing from its nostrils. [36]

CELEBES (LAKES OF), GREAT SUNDA ISLANDS, INDONESIA.

An undescribed species of crocodile is suspected to inhabit these lakes. [37]

CHACO, PARAGUAY

A strange creature is said to exist here. Described as a slug-like snake as broad as a horse, with the head of a dog and a barbed spike on a stumpy tail. Bernard Heuvelmans and Karl Shuker have suggested a giant catfish as the probable
identity of the creature [21,37]

CHAD LAKE, NIGER/NIGERIA/CHAD/CAMEROON, AFRICA

140 x 90 miles. Area: 6000 sq mi. Max depth: 23 ft. This vast lake Bernard Heuvelmans believes the lake is home to an unknown species of sirenian [37]

CHAMPLAIN LAKE, NEW YORK/QUEBEC, USA/CANADA.

109 x 0.5 to 11 miles. Area: 435 sq mi. Max depth: 399 ft. The monster called 'Champ' is probably America's most famous lake monster, reported hundreds of times. Suggested identities range from a surviving plesiosaur or zeuglodont to the highly unlikely tanystropheus. [7,10,38]

CHRISTINA LAKE, ALBERTA, CANADA

This lake is said to have produced many reports during the 1980s of a lake monster given the name of 'Christina'. Its general description is on a creature with a horse-like head, eyes the size of saucers - and, amazingly, hair said to be similar to Bo Derek's in the film *10* ! [30,39]

CHUZENJI LAKE, HUNSHU, JAPAN

Length: 7 miles. Reported home of a lake monster. Strange waves are said to break on the lake's shores, said to be caused by the monster's movements. [1]

CLADDAGHDUFF LOUGH, GALWAY, IRELAND.

Produced a sighting in 1956 of what was first believed to be a bullock, but on closer observation by the witness was seen to be an eel-like creature showing at least 10 ft of its body as it turned over, and revealing a white underside. [2]

CLEARWATER RIVER, ALBERTA, CANADA.

Supposedly in 1946 Robert Forbes observed a 20 ft fiery-eyed grey beast with pointed teeth and a scaly body rise up out of the water and snatch a calf from the bank of the river, which it then dragged under [30]

COMO LAKE, LOMBARDY, ITALY.

In 1946 fishermen were said to have observed a 'marine monster' 4 m long and covered with scales and red marks. [41]

CONGO RIVER, AFRICA.

Before 1890, J R Werner reported, a number of times, observing crocodiles up to 50 ft long and 4 ft thick, from a steamer he was travelling on. [40]

COWICHAN LAKE, VANCOUVER ISLAND, BRITISH COLUMBIA, CANADA.

Said home of a lake monster with the local Indian name of 'Tsinquan'.

CRESCENT LAKE, NEWFOUNDLAND, CANADA.

In 1960 four loggers reported seeing a giant conger eel at least 3 m long burrowing its way through a sandbank. [41]

CROLAN (GOWLAN) LOUGH, GALWAY, IRELAND.

This lough, which is connected to Lough Derrylea, has produced a couple of reports from the 1967s and 70s of a lake monster. There is also a rumour that a creature was trapped in a culvert, and subsequently died, in the 1880s. [3,15,16]

CULLAUN LOUGH, IRELAND.

I have found only one report from this lough - from Tony 'Doc' Shiels, the well-known magician, artist, author and self-confessed hoaxer. He reports seeing a single 4 ft long dark hump gliding through the water late one night. [32]

CUYAHOGA RIVER, OHIO, USA

In 1944 a number of people reported seeing an 18 ft long snake in the area, which reared up as high as a man, and had a taste for chickens [42]

CYNWCH LAKE, GWYNEDD, WALES.

A Wyvern (a two-legged dragon) was said to bask on the shore of this lake and move across the country in search of food. Interestingly, its body was said to have formed humps as it moved. [5,28]

DAKATAUA (DATAKAUJ) LAKE, NEW BRITAIN, BISMARK ARCHIPELAGO.

1400 ft x 30 ft. A horseshoe-shaped lake near to the sea, said to have no fish. This lake recently became famous in the cryptozoological world, because of what, at the time, seemed excellent video footage taken of a large unknown aquatic animal in the lake.

Recently, however, in an excellent article, Darren Naish has almost conclusively proved that the creature filmed was actually an indopacific crocodile. Whether the crocodile explanation explains all the sightings over the years is open to question. The creature has a number of local names such as 'Migo', 'Migaua', 'Massali', 'Mussal' and 'Rui'. [7,37,43,44,45,46,47]

DERG LOUGH, DONEGAL & GALWAY, IRELAND.

Although the legendary Irish hero Finn was said to have killed the monster in this lough, Captain Lionel Leslie was told in the 1960s by an old man that there was still a water horse in the lough. [1,5]

DERRYLEA LOUGH, GALWAY, IRELAND.

This lough is connected, as already mentioned, to Lough Crolan, and like that lough, many residents say that in the 1880s a giant eel-like creature was trapped in the 300 ft gully which separates the two loughs. [2,3,15]

D'ESPOIR BAY (SWANGER COVE), NEWFOUNDLAND, CANADA.

A most unusual lake monster was reported from here in 1952. It was described as approx 1 ft long, similar to a lobster but without the jointed tail. It had fish-like eyes, three pairs of legs and 3 inch long pincers. Despite only being reported twice, it has the local name of 'Maggot'. [41]

DEVILS LAKE, USA

A couple of reports from the 1890s were made of two reptile-like monsters which rose out of the water and had two fin-like paddles attached to the body. [31]

DHU LOUGH, GALWAY, IRELAND.

One report here by three witnesses of a creature with a head and mane like a horse, which rose up from the water. [3]

DOCHFOUR LOCH, INVERNESS, SCOTLAND.

Connected to Loch Ness. Here in 1995 T Regall claimed to have made sonar contact with a monster (Nessie venturing out?) after being led to the spot by a friend using a pendulum and a map of the loch. [i.e. map dowsing]

DOUBS RIVER, SWITZERLAND.

Length: 265 miles. A monster was reported from here in 1934, described as having an oval and blue back, small head on a long neck, and a yellow stomach, and it was said to undulate. [41]

DUBH LOUGH, GALWAY, IRELAND.

A number of reports were received from here in the 1950s and 60s. The most peculiar was a report from March 1962 when an A Mullancy and his son, whilst fishing, almost hauled a strange creature onto the shore. It was grey, had thick legs, a hippopotamus-like face, small ears, and was as big as a cow or ass. Most amazingly of all, however, it had a white pointed horn on its snout. It attempted to attack the boy, who fled with his father. The creature has not been reported since. [1,3,12]

DUBHRACHAN LOCH, SLEAT, SCOTLAND.

In 1870 the local Laird of Sleat unsuccessfully attempted to have the loch dragged, to look for a monster reported on the shore. It was at first believed to be a dead cow until it moved and swum out into the loch. [1,35,48]

DUCHENE LAKE (an enlargement of the Ottawa River), CANADA.

A couple of reports from here, from the 1870s and 80s, of a dark green serpent-like creature the size of a small telegraph pole. [1]

DUOBUZHE LAKE, TIBET

In 1972 Chinese soldiers were said to have shot and killed a strange animal in the lake, and dragged it to the local village. It was reportedly ox-like with legs like a turtle, skin like a hippopotamus, and short curly horns. What became of the creature, if it ever existed, is not known. [51,53]

DUVAT LOCH, ISLE OF ERISKAY, SCOTLAND.

In 1893 E Macmillan, while looking from a strayed horse, reportedly got within 60 ft of a strange creature which at first he presumed was his own animal. He realised the creature was bigger than his, and it then gave out a hideous scream. He said he could not get a clear view of it because of the haze present at the time. He then fled the scene. Was the creature connected with the Loch? We will probably never know for sure. [1,5,15]

DYSYNNI RIVER, WALES.

A veterinary surgeon told investigator F W Holliday that he saw the backs of two huge eel-shaped creatures moving upstream while the river was in spate. [5,15]

EAST COAST RIVER, TRINIDAD.

Said home to a 25 to 30 ft long undulating scaly serpent, known as the 'Huilla'. [37]

ECHO LAKES, TASMANIA, AUSTRALIA.

Alleged sightings of Bunyips here, in the 19th century. [29]

EDWARD LAKE, ZAIRE, AFRICA.

Natives say a creature, which they call the 'Irizima', lives here. They describe it as a marsh monster with a hippo's legs and an elephant's trunk (the long neck?), a lizard's head and an aardvark's tail. In 1927 an expedition visited the lake in an attempt to resolve the creature's identity but was unsuccessful. [7]

EIL LOCH, SCOTLAND.

In a book called 'September Road to Caithness' by an author only identified by the initials B B, published in 1962, the author describes observing 150 ft away a 3 ft black shiny object which resembled a blunt blind head of an enormous work, which made a disturbance and sank back into the water.

This is the only sighting I could find from this loch, and the lack of identity of the witness must make it suspect. However, the loch is an extension of Loch Linnhe, which has, in the past, produced a number of reports of sea serpent . [3,49]

ELLESMERE LAKE, NEW ZEALAND.

Sightings of an otter-like creature, 'Waitoreke', from the mountainous country around the lake. [1]

ELSINORE LAKE, CALIFORNIA, USA

This lake, which periodically dries up, has reports of a lake monster dating back to the 1880s. A report from 1970 described a snake-like creature 12 ft long and 3 ft thick, undulating as it moved through the water [28]

ERCEK LAKE, TURKEY

Has reports of a lake monster described as like a white aquatic horse. [53]

ERIE LAKE, OHIO-ONTARIO, USA-CANADA

The fourth largest of the Great Lakes, 9940 sq mi, linked to Lake Ontario. This lake has sightings of a monster dating back to the early part of the 19th century. The creature recently attracted media attention when the monster, known locally as 'South Bay Bessie', was seen several times. It was described as 3- - 50 ft long, humped, black or blue in colour, and with a head like a snake. [1,31,50]

ESQUEL REGION (mountain lake in), PATAGONIA

Was the site of an expedition in 1922, looking for what they believed was a plesiosaur, allegedly seen earlier.

EUROA (swamp near), AUSTRALIA

In 1890 a 27 ft long creature with a head like a bullgod and a tail as thick as a man's thigh was allegedly seen here. Apparently Melbourne Zoological Gardens used 40 men to comb the area for it. [29]

FADDA LOUGH, GALWAY, IRELAND

1.25 mi x 1800 ft. Depth approx 32 ft. This lough produced a sighting in 1954, witnessed by four observers, of a black creature with two 2 ft humps, a head held in a curve 3 ft out of the water, with its mouth open. Its skin was smooth and its tail forked. It approached within 60 ft and then swung around and dived.

In 1966 and 67 two unsuccessful attempts were made to flush out the monster. [1,2,3,15,16]

FIERY CREEK, VICTORIA, AUSTRALIA

Supposed sight of the killing of a bunyip by Aboriginals during or before the 19th century. A 9 m long outline was traced on the ground where the creature lay. The Aboriginals used to retrace and clear the outline, but eventually it was neglected and all that remains is a sketch drawn in 1867. [13,29]

FLATHEAD LAKE, MONTANA, USA

25 x 15 mi. This lake, fed by glacial meltwater, has had a number of sightings, usually described as a 25-30 ft black creature, sometimes with a fin protruding out of the water. Some of the sightings could be of a sturgeon, as they exist in the lake. However, some witnesses say the creature, known locally as 'Flattie', undulates, and it has been suggested that it could be a surviving zeuglodont. [1,32]

FLY RIVER BASIN, NEW GUINEA

Home to a possible undiscovered species of freshwater stingray known only so far by photographs. [43,52]

FOWLER LAKE, USA

Had reports in the 1870s and 80s of a lake monster: some said it was "the grandfather of all fish" and some believed it was an otter or a surviving beaver which used to live in the lake. [31]

FYNE LOCH, ARGYLL, SCOTLAND

One report here from the brother of author Gavin Maxwell: whilst driving by the loch he saw what at first he believed o be a sandbank. When he reversed to the spot, the object had gone. [3]

GEAL LOUGH, KERRY, IRELAND

Said home of a legendary creature called the 'carbuncle'. Pearl shells were said to be often found by the locals, and were believed to have come off the creature, which was described as a kind of snake covered also in gold and precious stones - and was said to make the lake shine at night. [1]

GENEVA LAKE, USA

In 1982 three witnesses were said to have seen a 100 ft long serpent rise 10 ft out of the water, with fierce eyes, hooked teeth, and scales with glistened in the sun. It turned and threw itself out of the water. Gary S Mangiacopra, who has investigated the case, believes the report is a hoax. [31]

GEORGE LAKE, New South Wales, Australia

Said home of a bunyip, although a seal has been suggested as an identity. [1,43]

GLAISH (GLASS) LOCH, SCOTLAND

Length: 3 miles. Around 1730 it was reported that religious ceremonies were performed to clean the waters, where a water horse was said to have been observed. [18]

GLASLYN LAKE (LLYN FFYNONLAS), GWYNEDD, WALES

Has a legend dating back to before the 17th century that a monster which had flooded the Conway Valley was dragged by oxen to the lake. Perhaps there is some truth in this, as in the 1930s two witnesses where climbing Snowdon reported seeing in the lake a grey line rise to the surface and then a pale head emerge. They were certain it was not an otter. [5,15,18]

GLENDARRY LOUGH (SRAHEENS LOUGH), MAYO, IRELAND

Breadth: 400 ft. Circumference: 1200 ft. Said to be extremely deep, it has been suggested that it is the crater of an extinct volcano. Scene of a faked photograph, the monster seems to spend more time cavorting around on land than in the water.

Sometimes there were classic Nessie-type sightings: a long neck with a snake-like head. But most sightings were of a creature 8-12 ft long, dark brown or black, with a long neck, long tail, and with a sheep-like head, running on and. It was said to have left three-toed tracks. Could the sightings have been of otters, their size being exaggerated? Or even sightings of the legendary Master Otter or 'Dobhar-Chu', as suggested recently by Karl Shuker. [1,2,53]

GOOSE CREEK LAGOON, USA

Nature writer H Sass and his wife, while boating, saw an object moving under water; and, with the aid of an oar, lifted part of it out of the water. It was bright salmon pink and orange, and as thick as a man's thigh. It had a smooth tail, short legs like an alligator, and was 5-6 ft in length. It slipped off the oar and back into the water. Sass believes it was a Giant Hellbender. [7]

GRANEY LOUGH, CLARE, IRELAND

A lake monster was supposedly seen every seven years; and an anonymous witness told of a swimming man being chased to shore by a giant eel. [1,2]

GREAT LAKE, TASMANIA, AUSTRALIA

Area: 44 sq mi. This lake is a lofty 3380 ft above sea level and had a report, from 1863, of a sighting of what was believed to be a bunyip. The witness was so close he could have touched it with his oar. It was the size of a sheep dog and had two small flippers or wings. It swam away and did not appear to dive. [29]

GREAT SANDY LAKE, MINNESOTA, USA

A single report here from 1886 of a hunter shooting at a monster that supposedly reared out of the lake. [1]

GUAVIARE RIVER, SOUTH AMERICA

Marquis de Waurin was told by natives that, often during periods of flooding, anacondas over 50 ft in length were seen, as thick as a canoe. [6]

HANAS LAKE, XINJIANG, CHINA

Length: 15 miles. Average depth: 500 ft. This lake became famous in the 1980s for the many sightings by observers (including biologist Professor Xiang Lihao) of a number of reddish coloured salmon-like fish with a length of approximately 33 ft. Investigations discovered that the fish had been reported for decades by the local people. [21,54,55]

HASLAR LAKE, ENGLAND

In September 1987 reports were received of a 12 ft eel which was said to have attacked swimmers and divers. The lake was drained and searched, but no evidence was found. The lake is situated very close to the sea, and, in my opinion, it would be possible for an eel to have existed in the lake, and migrated overland back to the sea. [56,57,58,59]

HERON (HERREON) LAKE, NEW ZEALAND

In the 1860s a Waitoreke (a supposed otter-like creature) was reported by two sheep farmers. It was described as dark brown, and the size of a rabbit. The two farmers promptly whipped it - and, not surprisingly, it yelped and disappeared into the water [1]

HOAN KIEM LAKE, VIETNAM.

In December 1996 a legend dating back to the mid fifteenth century of a giant turtle which was said to exist in the lake was proved to be a reality with the appearance of a very old and very large turtle with a shell length of a metre. Its green and yellow head was said to be the size of a football and it approached within six and a half feet of the shore proving that legendary monsters do sometimes become reality. [34]

HOWICK, KWA-ZULU NATAL PROVINCE, REPUBLIC OF SOUTH AFRICA

In October 1995 B Teeney said that he saw a twenty metre long creature at a waterfall near the town. He said that he would soon reveal evidence proving the existence of the monster but at the moment the world is still waiting. [23]

IKEDA LAKE, KYUSHU PROVINCE, JAPAN.

Area 11 square km. The local lake monster here has been dubbed `Issie` by the inhabitants of the area and it has been observed sporadically since the 1940s. It is described as being up to thirty feet long and is sometimes humped. Video film and photographs have been obtained of the creature. There are eels in the lake which have achieved lengths of up to two metres so the possibility of a giant eel being behind the reports cannot be ruled out. [7,10,60,61]

ILIAMNA LAKE, ALASKA, USA.

The largest lake in Alaska has produced many sightings over the years, usually of a whale or seal-like creature with a fin on its back, and which has sometimes been reported as squirting water into the air Explanations vary from a lost whale which has strayed from the ocean to a huge sturgeon or a species of freshwater seal. [8] [62]

INAGH LOUGH, CONNEMARA, EIRE.

In 1897, N. Colgan was told of a sighting of a water horse which emerged from the lough, shook its mane and then plunged back into the water. [1]

ITURI RIVER, AFRICA.

Length: 620 miles. In 1912 Alex Godart was told that a creature called the `nyama` resided in the river. It was said to kill men and eat their brains. It was as big as a hippopotamus, had a little head with feathers and a crest like that of a cockerel. [6]

JORDAN RIVER, TASMANIA, AUSTRALIA.

Supposed reports of a Bunyip were received from here during the 19th Century. [29]

KANAHWA RIVER, WEST VIRGINIA, USA

In December 1933 it was reported that two men captured a living octopus in the lake after its tentacles appeared over the side of their rowing boat. Its body was brought ashore and it measured three feet from head to tentacle. [7]

KASAI RIVER, CONGO, AFRICA

In the 1930s reports were produced of giant crocodiles seen in the river . [40]

KHAIYR (KHAYYIR) LAKE, YAKUTIA, NORTHERN SIBERIA.

Length: 650 yards. Depth: 30 feet. This lake has produced a number of sightings. Of particular note was a 1964 sighting by a biologist named Nikolai Gladkika of a long necked creature with a huge bluish-black body, two pairs of ill-defined limbs, a long tail, and a low triangular dorsal fin. It was standing on the shore on the opposite shore of the lake from him. There was also a sighting in 1942 by two Soviet pilots, who whilst surveying the lake observed two giant creatures which they described as being like giant newts. [1] [7] [12]

KLEIFARVATN LAKE, ICELAND.

This lake which is rich in fish and bird life produced an unusual sighting in November 1984. Two witnesses claimed that from a distance of a hundred yards they saw two animals leave the water. They were said to move like dogs but to be larger than horses. They returned to the lake where they swam like seals (presumably undulating). They left prints on the shore which resembled those of cows but which had three impressions. An unusual report, perhaps to be compared with the sighting from Duvat Lough in Scotland and Glendarry Lough in Ireland. [21] [62]

KOL-KOL LAKE, DZHAMBUL, KAZAKHSTAN, FORMER SOVIET UNION.

Although I have only been able to discover a single sighting from this lake of a one humped creature, forty five feet long and with a long neck which was reported in 1977 the `creature` does have a local name - "The Aidakhar". Recently it was claimed that the mystery had been solved by an expedition who said that currents washed away the mud causing whirlpools. They also claimed that the trumpeting sound attributed to the monster was caused by air being sucked into cracks connecting the lake with underground caverns. [21]

KOUKOUROU RIVER, CENTRAL AFRICAN REPUBLIC.

In the early part of this century a tribesman called Moussa said that he had observed a cat-like creature with the local name of "Mourou N'gou" which he described as being larger than a lion (approximately twelve feet in length) with a background pelage like that of a leopard, but additionally adorned with stripes. It emerged from the river close to a soldier in a canoe, seized him and pulled him down into the water causing the other soldiers in the detachment to refuse to cross the water at that point. Karl Shuker and Bernard Heuvelmans have both speculated that the creature responsible could be a surviving sabre toothed cat which has adapted to an aquatic lifestyle. [6] [65]

KUTCHARO/KOTCHARO/KUSSHARO LAKE, JAPAN.

Although supposedly having the local name of "Kushie" or "Kusshi" the only report that I can find from this lake is that of Professor S.Kirby who in 1980 said that a monster had been seen here on a number of occasions and even photographed. [10]

LABYNKAR LAKE, YAKUTIA, SIBERIA, RUSSIA.

Length: Approximately seven miles. Depth: 150-200 feet. There have been a number of sightings reported from this lake since the 1950's. What seems more amazing than usual about this monster is the number of times that it has been reported by qualified people. A geologist saw it in 1954. A team of geologists also reported seeing it snatch a bird from the air in the 1950s. It was reportedly seen through the ice by yet another team of geologists, and two expeditions (including trained zoologists) were sent to look for it in 1963 and 1964. Suggested identities vary with some biologists believing it to be a giant northern pike - up to nine foot in length. [1] [7] [12] [64]

LAGO LACAR

Over a hundred fathoms deep, this lake has reports of a lake monster with the local name of "Cuero" which means `cowhide` - a reference to the resemblance of the supposed animal to the hide of a cow. It is also known as `El Bien Peinado` which means `the smooth headed one`. There are reports of strange tracks on the shore and remains of what are supposedly the monster's meals - skins, furs and feathers have also been found. [1] [36]

LACHLAN RIVER, NEW SOUTH WALES, AUSTRALIA

In 1847 a shepherd boy reported seeing a Bunyip in a reed bank on the river. He described it as being as big as a calf, dark brown in colour, having a long neck, large ears and a long pointed head. It also had a thick mane of hair flowing down its neck and two large tusks. The boy ran away and so did the creature by way of a shambling gallop. As it disappeared the shepherd boy observed that the creature was possessed of a long tail. Tracks were later found that were said to be similar to a man's head [29]

LAKAGH LAKE, KERRY, IRELAND.

Area: five acres. There have been a couple of sightings from this lake. The first was at some time during the 1960's - a creature's head with two stumpy horns, several feet of neck and a long snake-like body was observed near a clump of reeds. In 1967 a Mr Wood observed a seven foot long yellow-brown object surface and then submerge a few yards away. [3] [15] [16]

LAGARFLOT LAKE, ICELAND.

This narrow mountain lake which is fifteen miles long has had reports of a monster called the 'skrimsl' which date back to 1345. Reports were received during the eighteenth and nineteenth centuries of a creature described as having a seal-like head, a six foot long neck, a 22 ft body and an eighteen foot tail. A reported carcass of the creature was washed up on the shore during the nineteenth century. It was described as a mass of bones and flesh and was different from the carcass of a whale. [1] [12] [48]

LAURISTON RESERVOIR, AUSTRALIA.

There was a report in 1949 of a bunyip described as being 4 feet long with long, shaggy ears. [29]

LEMAN LAC, SWITZERLAND.

Although there are rumours of a monster here, German Cryptozoologist Ulrich Magin believes that they are a hoax [42]

LETHBRIDGE IRRIGATION SYSTEM (RESERVOIR), ALBERTA, CANADA.

There was a report here in 1945 of a 12-14 foot long creature swimming with its head several feet out of the water [30]

LEURBOST (INLAND LOCH), ISLE OF LEWIS, SCOTLAND.

There was a number of reports from here in the eighteenth and nineteenth centuries. One was a rumour of a huge conger eel captured during the 1770s. Another was a report from 1856 of a monster observed by a crowd of people and which was described as looking like a large `peat stack` with huge fins and the form of an eel [5]

LICKING RIVER, COVINGTON, KENTUCKY, USA.

There is a single report from 1959 of an octopus-like creature with tentacles and a lopsided crest which allegedly surfaced and emerged onto the bank. [7]

LIKOUALA RIVER, AFRICA

In the 1960s Nicholas Mondongo was standing on the river bank when he observed a huge animal surface. It had a long slender neck and a well defined head, a bulky elephant-like body and four massive legs. It also produced a long, tapering tail. He observed the creature for three minutes from a distance of only fifteen yards. Its overall length was thirty feet and the creature eventually submerged.

In 1981 The Roy P. Mackal expedition to search for the `Mokele-Mbembe` was completing a curve on this river when they observed a six inch high wake from a creature which had just submerged. Crocodiles were ruled out and it is said that hippopotamuses do not exist on this river. Nearby the expedition also found a possible trail of a `Mokele-Mbembe`. [7] [67]

LISMORE LAGOON NORTH OF), NEW SOUTH WALES, AUSTRALIA.

During the early 1970s there were a number of sightings of an unusual animal which was described as being hairy with a head like a dog. It was said to be six feet in length with the girth of a twelve gallon drum. It was said to feed on wildfowl and was also rumoured to be behind the mysterious disappearance of two aborigines. [29]

LITTLE MIAMI RIVER, OHIO, USA

Rumours of giant frog-like creatures here date back to the 1950s but became prominent in 1972 when two police officers encountered a frog-faced animal the size of a dog which was climbing over a guard-rail between the road and the river. A later report by two boys from 1985, said that whilst they were skimming stones across the surface of the water they sighted a very large frog which jumped and was reportedly four feet wide. [21] [41]

LITTLE MURRAY RIVER, AUSTRALIA

During the late 1940s three witnesses reportedly saw a strange animal on two occasions. The first time they saw it lying on the river bank. It was between three and three and a half feet long, black in colour and at first they thought that it was a pig. The second sighting was at the same spot a month later. The creature was swimming upstream, its head and neck were nine inches thick and stuck a foot out of the water. They said that it was spouting water five feet into the air. It swam to the opposite bank where it lay in shelter. It possessed two bright eyes and emitted a loud, shrill whistle. [29]

LOCH LOCHY, INVERNESS, SCOTLAND.

Length: Ten miles. Average width: three fifths of a mile. Maximum Depth: 531 feet.

There are a number of reports from this loch of a monster that has been given the nickname of `Lizzie`. These date back to the late 1920s. There is even a photograph which was taken in 1937 showing a strange object surfacing in the Loch. Maurice Burton, however, believes this to have been a vegetable mat. It is, however still open to question. The most recent sighting was in September 1996 when over sixteen people observed a creature the size of a dolphin with three humps which moved from side to side. [1] [3] [12] [35] [36] [68]

LOCH LOMOND, CENTRAL AND STRATHCLYDE, SCOTLAND.

Length: 22 miles. Breadth: Up to five miles. maximum depth: 623 feet

This, the largest lake in Scotland, has produced a few reports over the years. In 1653 Blaeus Atlas said that the loch had "waves without wind, fish without fins and a floating island". There were a couple of reports during the 1960s and a report from 1980 of a five foot long head and neck which appeared followed by a long curved shape. [1] [3] [12] [18] [32] [36] [69] [70]

Loch Lomond

LONG POND, NEWFOUNDLAND, CANADA.

In 1967 there was a report of an eel like monster a foot wide and thirty feet long with a salmon like head and a trout-like tail. [41]

LYNN, (NEAR) INDIANA, USA.

In 1960 Dan Craig reportedly saw a strange creature in a well on his farm. It was said to have a domed head, two bulbous eyes and eight tentacles each as long as a man's arm. The well was eventually drained and a boy climbed down - he said that he had observed a similar creature. A few days later however, a conservation officer hauled from the well a flesh coloured ball of sponge and a segment of garden hose! As can be imagined the creature was never reported again. [7]

MAAS RIVER, BELGIUM

Length: 560 miles. In 1979 a three foot long crocodile was seen in the river near the village of Ombert. It soon earned the local name of 'Maasie'. A search by the police revealed nothing although a herpetologist at Amesterdam Zoo speculated that the creature may have been a pet alligator which had been released into the river. [41]

MACQUARIE RIVER, AUSTRALIA.

After the Second World War, Jack Mitchell researched a number of reports of a Bunyip seen along a 200 km stretch of river between Wellington and Warren. It reportedly had a head like a calf, made a fearful noise and possessed tremendous strength, easily breaking through the fishing nets of the local population. It was sometimes seen sunning itself on the banks of the river or swimming against the current whilst thrashing the water. [29]

MAGDALENA RIVER, COLUMBIA.

In 1921 a sighting, supposedly of an Iguanadon was made here. [7]

MAGGIORE LAKE, ITALY/SWITZERLAND

Length Thirty nine miles. Breadth: Five and a half miles. Area: Eighty two square miles. The monster of this lake was supposedly mentioned in a travel book written by the novelist Stendahl in the 19th Century. recently, however, French cryptozoologist J.J.Barloy has tried and failed to locate this reference. The creature was, however, reported again by two fishermen in 1934 who saw the creature near the mouth of the River Ticino [1] [41]

MAIN RIVER, GERMANY.

Length 320 miles. In August 1983 a witness reported seeing a giant serpent swimming in the river near Frankfurt. Later, water police found the body of a seven foot snake which was taken to the Senkenburg Institute for identification.

MAJOR LOUGH, MONAGHAN, EIRE.

In 1963 three youths returning from a fishing trip reported seeing a monster in the lake. They described it as splashing up and down like a sea lion, and having a hairy head and two horns. They threw stones at it but fled when it made towards them. Paddy Brady, a local resident, also reported having seen the creature. [1]

MALERN LAKE, SWEDEN

In 1765 the people of Stockholm reportedly tried to dissuade the Bishop of Avranches from swimming in the lake because of a dragon called `The Necker` which they believed lived there. [1]

MALMSBURY RESERVOIR, VICTORIA, AUSTRALIA.

This reservoir is only two km from Lauriston Reservoir and in the 1870s produced a report of a large and very dark dog-like creature with a head resembling that of a seal which then dived and disappeared. A similar animal was observed on a later occasion.

MAMFE POOL, UPPER CROSS RIVER, CAMEROON, AFRICA

In 1932 the famous fortean zoologist Ivan T Sanderson and fellow zoologist Gerald Russell were paddling their canoes at sundown when they heard a terrible roar emitting from a cave. They moved nearer, heard a second roar, and then something enormous rose out of the water, roared and then plunged back into the depths. They believed that it was the head of a huge creature, shiny black in colour and shaped like a seal, but flattened from above to below and the size of a full grown hippopotamus. [26]

MANITOBA LAKE, MANITOBA, CANADA.

Area: 1,817 Square Miles. This huge lake has produced a number of reports dating back to the early part of the 20th Century. These animals usually have the classic `Nessie` type appearance, of a single hump or a long neck. Roars have allegedly been heard and there are reports of more than one creature having been seen at a time. There is even a rather convincing photograph which was taken of the creature in 1962. [1] [10]

MARAKOPA RIVER, WELLINGTON, NEW ZEALAND.

A rather aggressive monster resembling an archaeocete is rumoured to exist in this river. [32]

MARA RIVER, TANZANIA, AFRICA.

Reportedly, a hunter observed an animal floating down the river on a large log. It was approximately sixteen feet long, spotted like a leopard, covered in scales and had a head like that of an otter. he fired at it, apparently hitting it, as it slid into the water. [6] [7]

MAGGORI RIVER, AFRICA.

Apparently the hunter John A.Jordan observed a creature known locally as the `Dingonek` here. He described it as being between sixteen and eighteen feet in length, with a massive head and tusks like those of a walrus which projected from its upper jaw. It had a broad tail, a wide spotted back and a scaly body like that of an armadillo. Bernard Heuvelmans and Karl Shuker have both hypothesised that this may have been an aquatic sabre toothed cat. [6] [7]

MARMORE RIVER, MATTO GROSSO, SOUTH AMERICA.

In 1931 Harold Westin was travelling down the river when he observed a strange animal on the shore. It was a twenty foot long dinosaur-like creature, greyish in colour with an alligator-like head and four lizard-like feet. Its eyes glowed scarlet in colour. Westin shot at it, but although he apparently hit it, the creature remained unharmed. [7]

MASK LOUGH, GALWAY, IRELAND.

There is a legend here which dates back to 1674 of a monster which came onto the shore and dragged a man into the water. Fortunately he fought it off, but soon after a strange carcass was found rotting in a cave. A more recent report is from the 1960s and tells of a head and tail appearing followed by a five to six foot long hump very much like the back of an eel. [1] [3] [15]

MASSAWIPPI LAKE, CANADA.

Depth: Over 1200 feet. There is a legend associated with this lake telling of children being warned not to swim in the water because of the monster which the locals say is a fish like creature with the head of a cow. [19]

MEDJERDA RIVER, TUNISIA.

This is the site of the most ancient monster record that I have been able to find. In 255 BC Roman troops reportedly used catapults to battle and subdue a 120 foot serpent. [42]

MEIKLIE LOCH, SCOTLAND

This lake is home to a legend of water horses and bulls. [12]

MEKONG RIVER, VIETNAM.

Length: 2,750 miles. In 1969 there were rumours of a nine nostrilled (how did they know? Did they count them?) water monster seen in the area. [72]

MEMPHREMAGOG LAKE, QUEBEC, CANADA/VERMONT, USA.

Length: 32 miles. Until 1992 there had been nearly 150 sightings of this particular lake monster. These date back to the early 1800s. It has the local name of 'Memphre'. A typical sighting is that of August 7th 1992, when Jennifer Malloy and her husband, nephew and niece were travelling in a boat when they saw the creature. Approximately two hundred feet away a wake appeared and in the midst of it they saw a shiny, green hump which rose two feet above the water. The hump was approximately six feet long and moved with an undulating motion. It was observed for between 35 and 40 seconds. [1] [73] [74]

MENDOTA LAKE, WISCONSIN, USA.

Until the 1920s there were a number of reports of monsters from this lake. One from June 1883 was by a Mr B.Dunn and his wife who, whilst boating, observed an object which moved near to them. As it came closer they noticed that it was light greenish in colour and covered in white spots. It raised a snake like head and possessed a forked tongue. It hissed and attacked them. Dunn hit it with an oar. One of the fangs of the beast became embedded in it. Dunn then attacked the creature with a hatchet. Eventually it gave up and sank below the water leaving Dunn and his wife with the souvenir of the fangs embedded deep in the oar. He allegedly kept them. If the story is true, could the creature have been an escaped or out of place python or boa? All the reports were of a snake like creature and one was by students who claimed that the creature flicked at one of their feet with its tongue. [31] [82]

MICHIGAN LAKE, MICHIGAN, USA.

Area: 22,390 square miles. There is a single report from this huge lake which dates back to 1892. A party of tourists were travelling on the lake when they observed a 60-75 foot monster four feet in diameter. they saw it from some distance and was said to roll and tumble on the lake and would intermittently dive down and reappear. Its eyes were described as being the size of dinner plates and it was said to have immense jaws and sharp teeth. It was dark brown in colour and its body tapered like that of a snake. [31]

MIDGEON LAGOON, NEAR NARRANDERA, AUSTRALIA.

In 1872 witnesses reported seeing a bunyip here. It moved forwards them at a great speed making a loud noise. It was described as being half as big again as a retriever dog and was covered in shining black hair, which appeared to be about five inches long and which floated on the surface. It had well defined ears and was observed stationary for half an hour after which it swam away. [29]

NATCHEZ DISTRICT, MISSISSIPPI RIVER, USA.

Length: 2,350 miles. There have been reports of freshwater sharks having been seen in this river. [52]

MJOSA LAKE, NORWAY

Length: 62 miles. breadth; between one and nine miles. Maximum Depth: 1,000 feet.

There have been rumours of a monster here since the 16th Century. In 1884, one of the world's first cryptozoologists, A.C.Oudemans, was told of strange movements seen on the lake which were attributed by locals to the movements of a sea-serpent which was seen from time to time. [1]

MODEWARREL LAKE, VICTORIA, AUSTRALIA.

In the early 1800s William Buckley observed a bunyip in the lake on numerous occasions. He even unsuccessfully attempted to spear it. He only observed the back of the creature and he described it as being dark grey in colour and covered in feathers. It was the size of a full grown calf and appeared only when the weather was calm and the water still. [28] [29]

MOFFAT LAKE, CANADA.

Length: 3 miles. Maximum depth: 33 feet. There have been a few reports of a monster here dating back to the 1880s. A certain Mr Macrae even mentioned a report of a friend of his who had seen it on the shore. At first he thought that it was a big burnt log and he was about to step on it when it scuttled back into the lake. [19]

MOLONGLO RIVER, AUSTRALIA.

In 1886, a bunyip the size of a dog with the face of a child was reportedly seen in the river. Given the description of a child like face it may well have been a seal. [41]

MONONA RIVER, WISCONSIN, USA.

In 1897 there was a report of a monster at least twenty feet long which was observed in bright moonlight and which was travelling at high speed. It had the classic appearance of an upturned boat.

MORAR LOCH, INVERNESS, SCOTLAND.

Length: eleven miles. Breadth: Approximately one and a half miles. Maximum depth: 1,017 feet. This is undoubtedly Scotland's second most famous lake monster, and it has the local name of `MORAG' or `MHORAG'. There have been many sightings dating back as far as the 1880s, but the monster was brought to the attention of the general public in 1969 when it received wide media attention after allegedly colliding with a boat after which it was fended off with an oar. The animal survived the bullet which was fired at it and is usually described as being like an upturned boat or a long necked creature with a small head. [1] [3] [12] [48] [75] [76]

MUCK LOUGH, DONEGAL, IRELAND.

Length: Three Quarters of a mile. Breadth: half a mile. (Limited food supply - mainly Brown Trout). reportedly in 1885 a young woman who had waded out into the Lough to pull bog-bean saw a strange animal with large eyes making towards her through the water. She quickly moved to the shore. It was seen on a number of occasions over the next few months when usually two humps were observed. [1] [3] [12] [16]

MURRAY RIVER, NSW AUSTRALIA

In 1857, a naturalist called Stocqueler claimed to have observed from his boat a number of strange `fresh water seals`. They had two small paddles attached to their shoulders, a long swan-like neck, a head like a dog, and a curious bag hanging beneath the lower jaw like the pouch on a pelican. Their lengths varied from five to fifteen feet and their colour was a glossy black. It must be mentioned here that in the 19th Century seals were identified as far as four hundred miles up the river. [29]

MURRUMBIOGEE RIVER, NSW, AUSTRALIA.

In the 19th Century a couple of reports of a Bunyip were received from this river.. There was a sighting from the latter part of the century of a creature the size of a three month old calf basking on a sand bank by the water's edge. When it was disturbed it wriggled into the river. This sighting could well have been a seal. As with the sighting from the MURRAY RIVER (see above) a number of seals were reported from these waters during the nineteenth century. [29]

MYLLESJON LAKE, SWEDEN.

There have been rumours of many sightings from this lake over the years, and there is a local legend that an attempt was made to capture this animal using a large baited hook. The legend goes on to say that the monster was so strong that it managed to uproot the tree that the line had been tied to. [28]

NACORRA LOUGH, CO. MAYO, EIRE

Length: two to three miles. Before the first world war four witnesses reportedly saw the water disturbed by a huge black shape that rose and swam the length of the lough in a few minutes. Then other similar shapes appeared, diving like seals but appearing bigger than houses. A misidentification of a boat wake seems unlikely as these observations were made through binoculars. [30]

NAHOOIN LOUGH, GALWAY, IRELAND.

Length: 300 feet. Breadth: 24 feet. Depth: 22 feet. There was a sighting in the 1940s of a creature which rolled over showing a pale belly and was reportedly as wide as a car. In the late 1960s there were a number of other sightings and two attempts (one by Roy P. Mackal) were made to capture the creature. [2] [3] [7] [15] [16]

NAHUEL HUAPI LAKE, PATAGONIA, ARGENTINA, SOUTH AMERICA.

Length: 40 miles. Breadth: 6 miles. Area: 210 square miles depth: Over 1000 feet. There have been a few reports over the years of a monster with the local name of "Nahualito". A good sighting took place in January 1994 when twenty witnesses saw a thirty foot long grey-green creature with several humps. It made a lot of waves and a loud snorting noise. Suggested identities for the monster range from that of a plesiosaur to a species of duck called "Pato Vapor" which has a very reptilian appearance. [1] [7] [66]

NEGRO RIVER, BRAZIL.

The famous explorer Percy Fawcett reportedly shot and killed a giant anaconda on the river. He said that forty five feet of the body lay out of the river whilst another seventeen feet remained immersed. The body was supposedly not more than a foot in diameter. [6] [10]

LOCH NESS, INVERNESS, SCOTLAND

Length: 24 miles. Maximum Breadth: 1.96 miles. Depth: Over 750 feet. Area: 21.78 Square miles.

Where do you start? It is the world's most famous monster - there can hardly be anyone in the developed world who has not heard of "Nessie". She has had dozens of books and hundreds of articles written about her and her identity has been postulated as everything from a long necked seal or a primitive whale to a giant invertebrate or the ever popular plesiosaur. Deep down, I have always favoured the giant eel theory myself! There have been hundreds or sightings, usually of an 'upturned boat' like hump or a long neck protruding out of the water. It has inspired more searches than any other cryptid. The general public are under the impression that most of these searches have been well financed, where in fact the opposite has usually been the case.

The Loch has also had its full time investigators pioneered by the late Tim Dinsdale. These days Steve Feltham lives by the water's edge having given up his job to search full time for the monster and the best of luck to him! There have been many photographs and films taken of the monster (although many of these have been dismissed - hard luck Frank Searle). Perhaps the best photographs are those taken in 1972 by Robert Rines, and the film taken a decade or so previously by Tim Dinsdale. Sonar contacts have also been made over the years.

The search will probably continue for many years to come and even if nothing is ever found it has to be said that it was "Nessie" who started the whole metaphorical 'lake Monster Ball' rolling, and has probably inspired more cryptozoologists than has any other mystery beast. [1] [2] [5] [8] [10] [12] [15] [18] [28] [36] [38] [42] [43] [49] [78] [95]

NIAGARA RIVER, USA

There have been reports of a lake monster here dating back to 1817. [41]

NYASA LAKE (NOW LAKE MALAWI), MALAWI/MOZAMBIQUE, AFRICA

Length: 350 miles. In the early part of the 20th Century Arnold Drummond-Hay claimed to have seen a sea-serpent on the lake although it may well have been a large python swimming with its head erect. [1]

OCONOMOWOC LAKE, WISCONSIN, USA

There were rumours of a lake monster here in the late 19th Century although some believed that it was only a beaver. Gary S. Mangiocopra believes that a likely candidate for this creature is a fish that had achieved a gigantic size. [31]

OICH LOCH, INVERNESS, SCOTLAND

Length: Four miles. Breadth: Maximum 0.3 miles. Maximum Depth: 154 feet. This lake produced a number, of monster reports in the 1930s. One of the most interesting was from three men who observed a creature making its way across land from Loch Ness to Loch Oich. Another in 1936 was of a huge black snake-like body which had risen to the surface followed by the appearance of a head that was vaguely dog-like. The Loch was the subject of an unfortunate hoax by the Scottish Daily Express in 1961 who gave their fake monster the name of "Wee Oichy". [1] [12] [35]

LAKE OKANAGAN, BRITISH COLUMBIA, CANADA.

Length: 69 miles. Breadth: 2.5 miles. Area: 127 Square Miles. Depth: 800 feet.

Canada's most famous lake monster has been reported over two hundred times and has been given the popular name of 'Ogopogo' and the Native American name of 'Na ha ha itkh'. Like the creatures of loch Ness these animals have been videoed and photographed on a number of occasions. Unlike "Nessie" however there are hardly any reports of these creatures having a long neck. The reports of a serpent like creature which can achieve a length of up to seventy feet long and often sports a fluked horizontal tail suggest that 'Ogopogo' is a surviving primitive whale (possibly a zeuglodont). [1] [7] [12] [19] [28] [48] [49] [79] [80] [81] [82]

OKAUCHEE LAKE, WISCONSIN, USA

In the 1880s there were rumours of an immense fish existing in this lake. A fisherman at the time claimed to have seen a 'Buffalo Fish', six feet long, with an estimated weight of eighty to ninety pounds. It is likely that this sighting was the source for the rumours. [31]

OLENTAGY RIVER, OHIO, USA.

A strange creature was observed in this river during April 1982 by a number of witnesses who included fire officers and policemen. One officer believed that what they had seen was a hippopotamus. The local zoo was checked but none were missing. The local police stretched their imagination to unheard of extremes and christened the creature 'The Olentagy Monster'. It has been speculated that the creature may have been a large male Northern Elephant Seal. [67]

LAKE ONONDAGA, NY, USA.

In 1902 a fisherman netted a creature in the lake. It was identified by professor J Wilson as a squid. Shortly afterwards a second specimen was caught in the area - an area where salt springs are known to exist. Dr Karl Shuker has speculated that the water may be saline enough for an exclusively marine creature to exist there, and that they may have arrived in its post glacial history when the lake was still connected to the sea. [7]

LAKE ONTARIO, ONTARIO, CANADA

Although this is the smallest of the Great Lakes it is still a massive 7,540 square miles in area and is the reputed home to a lake monster which was reported as far back as 1829 when a hideous water snake or serpent of huge dimensions was seen. Also of interest is that author Farley Mowat and other passengers on a schooner claim to have witnessed the dorsal fin of a shark rising up out of the lake in the late 1960s. [1] [41]

ORANGE RIVER, RSA, AFRICA

Length: 1,300 miles. There were a number of sightings of a monster here between the 1890s and at least the 1950s. In 1910 six witnesses reported seeing a creature that was swimming rapidly against the current in the swirling rapids. It had a huge head and a ten foot long neck like a bending tree. One of the witnesses, F.C.Cornell believed that it may well have been a gigantic python. [1]

ORIGURE LAKE, BOLIVIA

A giant fish given to capsizing canoes is said to live here [37]

ORTOIRE RIVER, TRINIDAD

There have been reports over the years of a strange creature with the local name of 'The Huillia'. It is said to be twenty five to fifty feet long and it is described as undulating. It has a slender scaly body and can move as fast as a steam boat. It is said to come ashore and migrate overland from one stretch of water to another. Dr Karl Shuker has proposed that it may be a surviving zeuglodont. [7]

PAIKA LAKE, NSW, AUSTRALIA

In the mid 19th Century George Holder reported that some of his employees had seen a bunyip in the lake.

PARAGUACO RIVER, BRAZIL

reportedly in early 1995 a party of geology students were at the point on the river where it passes through Orobo and observed two dinosaur like creatures bathing in the shallows of the river. Each was about thirty feet long with a huge body, a fearsome head on a six foot long neck and an eight foot tail. [7]

PATTENGGANG LAKE, WESTERN JAVA

In the late 1970s reports were received of an eighteen foot long turtle or fish-like creature which was said to exist in the lake. The local fishermen treated it with great reverence burning opium to keep it pacified. [7] [21]

PAYETTE LAKE, IDAHO, USA

Length: 7 Miles. In the 1930s where were rumours of a monster here but it wasn`t until 1941 media attention was attracted when over a period of two months thirty people claimed to have seen what was variously described as a three humped or a long periscope-necked creature. It was given the local names of "Slimy Slim", "McCall Monster" and "Sharlie". [1]

PELLEGRINI LAKE, ARGENTINA

In the 1930s zoologist Hans Krieg mentioned reports of a `saurian` which was said to have been seen flying and swimming in the lake. Expeditions were reportedly sent to look for it. [66]

PERENE RIVER, SOUTH AMERICA

In 1946 Leonard Clark was told by several tribes of a gigantic plant-eating beast which resembled a sauropod. [7]

PERUGIA (MARSHES NEAR), ITALY

In the early 1930s an aquatic monster was said to have been seen in these marshes which interestingly enough, are connected to the sea by the River Tiber, the reported site of a sea serpent encounter in the 6th century. [41]

POHENGAMOK LAKE, QUEBEC, CANADA

Length: 7 miles. Breadth: Approximately one mile. This lake has produced many sightings which date back to the 1870s of a monster with the local name of `Ponik`. Most sightings are of a back often described either as being jagged or having a dorsal fin Professor Vadim Vladykov who has been interested in the monster since 1957 believes it to be a giant sturgeon. [10] [19]

PORT PHILIP, VICTORIA, AUSTRALIA

In the 19th Century a creature called the `Tuntapan` was said to exist here. It was as big as a bullock, it had an emu's head and neck, a horse's mane and tail and a seal's flippers. It was also said to lay turtle's eggs in a platypus nest. When it was tired of its diet of crayfish it was also said to eat Aborigines! [1] [6]

PO RIVER, (NEAR GORO), ITALY.

With a length of 415 miles this is the longest river in Italy and despite being heavily polluted it was the site of 1975 monster reports. The strange creature was described as being large and snake-like, but possessing legs. It was said to be about ten feet in length and to be as thick as a dog. Experts speculated that it was an escaped crocodilian. [41]

QUEANBEYAN RIVER, AUSTRALIA.

In the late 1800s John Gale was duck shooting when, approximately a hundred yards away from him, he saw a dog-like creature rise up out of the water and upon seeing him dive back down again. Other witnesses were also said to have seen the creature which may have been a seal. [29]

QUOICHLOCH, INVERNESS, SCOTLAND

This lake which has been turned into a reservoir is nine miles long with a maximum width of a mile and a half. During the 19th Century there were a couple of reports of a monster here. One was of the creature lying on the shore close to the water, and the other was of it swimming under water close to a fishing party who were travelling in a boat. [1] [12] [48]

RANNOCH LOCH, TAYSIDE, SCOTLAND.

It was rumoured in the 1700s that a water monster existed in this loch. Later, in the 1860s, it was said that local people would not venture near the bank of the loch without guns. [1] [17] [18]

RED CEDAR LAKE AND RIVER, WISCONSIN, USA.

A single report from here in 1891 described an undulating form "like a huge snake or fish" was seen. It had a huge head and a row of protuberances like a huge swath along its back. It has been suggested that it may have been a hoax, but if we disregard the unfish-like undulations, could it not have been a huge sturgeon? [1]

REE LOUGH, IRELAND

Length: 17 miles. Breadth: 7 miles. Maximum depth: 120 ft. There have been legends of a monster here since the seventh century A.D, but more recently there have been a number of reports from the 1950s and 1960s. The most famous of these sightings was that in May 1960 when three priests were fishing when they saw a creature about a hundred yards from them. It had a round hump about eighteen inches long, and two feet in front of it there was a serpent-like head attached to a neck between two and two and a half feet long. It was moving through the water and they watched it for between 2 and 3 minutes before it submerged. [1] [2] [3] [10] [12] [16] [49]

RHONE RIVER, FRANCE

Length: 505 miles. In the mid 1950s there were reports of a monster in the river, and in 1964 a sea serpent was seen near the river's mouth. [41]

ROCK LAKE, WISCONSIN, USA.

There were a number of reports during the nineteenth century of a monster. The earliest of these described a `saurian` lurking in a clump of reeds in either 1867 or 1875. Twenty years later two fishermen on the lake observed a creature moving through the water. As they got nearer a huge horse-like head with serpentine features reared out of the water followed by an arched back. [1] [82]

ROCKY LAKE, SOUTHERN AUSTRALIA, AUSTRALIA.

There is a single report dating from 1853. At night the witness, who was unable to sleep, heard a noise on the lake. He saw a blackish shape move towards the bank where it raised a large head on a horse-like neck out of the water. It possessed thick, bristly hair and kept near to the bank. As it moved down the river its length was estimated as being between fifteen and eighteen feet. [29]

ST FRANCIS`S LAKE, QUEBEC, CANADA.

Length: 21 miles. Maximum Depth: 178 feet. This lake, which is connected by river to Lake Aylmer has produced a number of sightings since the 1960s. One witness observed the creature for an hour and a half through a telescope. He described it as being between fifteen and twenty feet long, having a flat, elongated head, round eyes and a jagged back. It undulated as it moved. [19] [83]

ST JOHN'S RIVER, FLORIDA, USA

Length: 300 miles. Since the 1950s there have been reports of a strange pink creature which has been (not unsurprisingly) dubbed "pinky" by the less imaginative locals. The best sighting was in 1975 when five witnesses saw the man-sized head of a creature surface twenty feet away from them. It possessed a pair of horns and what appeared to be either gills or fins hanging down. Its eyes were dark and slanted and its neck protruded about three feet out of the water with what appeared to be a serrated upper surface. One witness likened it to "a dinosaur with its skin pulled back so that all its bones were showing", and not surprisingly it was pink in colour, Suggested identities have ranged from it being a surviving Thescelosaurus to an unknown species of giant salamander. [7]

SACRAMENTO RIVER, CALIFORNIA, USA.

In 1891 the head of a giant lizard was seen as it apparently emerged from the water [7]

SADDLELAKE, ALBERTA, CANADA.

In the 1980s reports were received of a lake monster with a head shaped like a horse, eyes the size of saucers and hair which looked like it was either beaded or in dreadlocks. [39]

SARAWAK/LAWAS RIVER, BORNEO.

There were a number of sightings during 1985 of a monster with the head of a cow, a neck the size of a forty gallon drum, and eyes like electric light bulbs! Unnamed experts identified it as a Dugong! [64]

SARY-CHALEK LAKE, TURKESTAN

In 1963 Moscow Radio reported stories of a monster in the lake, although Dr S.K.Klumov believed that the reports could have been attributed to a string of cormorants swimming across the surface of the water. [1]

SASKATCHEWAN RIVER, SASKATCHEWAN/ALBERTA, CANADA.

There have been reports of a monster here since the late 1940s. They describe a creature between five and eight feet long, covered in dark fur and with a head like an alligator. [30].

SEATTLE LAKE, WASHINGTON, USA

This is one of those true rarities - a lake whose monster mystery has been solved. For years the local population had told of a voracious monster which preyed upon the duck population of the lake. In November 1987, he body of an enormous sturgeon, eleven feet long which weighed 900 LB and was estimated to have died of old age at an estimated eighty years was found. [21]

SELJORD, (SELJOROVANNET), TELEMARK, NORWAY.

There have been reports of a monster in this lake for about three hundred years, and it has been given the local name of `Seljordsorm'. In 1963 A.J.Lindstol saw a creature similar to a horse, and in 1986 Aasmund Skori saw a bow appear - it was about four and a half feet long, as thick as a man`s thigh and black in colour. [19] [70]

SENTANI LAKE, NEW GUINEA

During World War 2, anthropologist George Agonino was attempting to obtain a supply of fish with the aid of a hand grenade when he inadvertently brought a twelve foot shark to the surface. He made a sketch of the creature. [37] [53]

SHANAKEEVER LOUGH, GALWAY, EIRE.

This is a small lough connected to Lough Auna and is only fifteen feet deep. There have, however, been a number of sightings of monsters here since the 1950s. The creature was supposedly seen on land in 1958 by A.P. Canning who described it as being like a black fowl. It was circling around a donkey but seemed higher than it. It had legs, a long neck and ears. It then disappeared back into the lough. Other reports are of a humps in the water, and occasionally there are reports of a giant eel. [2] [3] [15] [16]

SHANNON RIVER, IRELAND

At 240 miles long this is the longest river in Ireland. In 1922 many people on both sides of the river, as well as the crew of a ship upon the water witnessed a monster with a swan like neck at least twelve feet long. It had a smallish head which waved from side to side and bright shining eyes. Behind it was a cone shaped hump, ten to twelve feet in length which protruded a few feet out of the water.

It headed upstream at a slow speed. It seemed to notice the bridge in front of it at which it made a left turn and headed downstream towards the sea.[12]

SHIEL LOCH, HIGHLANDS, SCOTLAND.

Length: 17 miles. Maximum Width: One mile. Maximum depth: 420 feet.

This is Scotland's third most famous lake monster haunt - it was even mentioned on an episode of *The X Files* entitled "Quagmire". It has produced a number of reports dating back to the nineteenth century of a classic Nessie-like creature, with one sighting of the creature apparently on land. [12] [48] [84]

SHUSWAP LAKE, BRITISH COLUMBIA, CANADA

There have been a number of reports over the years of an animal which has been given the local name of 'Sicopogo' or 'Shuswaggi' and the Native American name of 'Ta-Zam-A'. A sighting in 1984 by the Griffiths family was of a creature three hundred yards away, which when observed through binoculars was between twenty and twenty five feet long, had seven grey humps, moved in a straight line and looked like a snake. [21] [51]

SLAGNASSJON LAKE, BLEKINGE, SWEDEN

This very deep lake produced a strange encounter in 1965. The two witnesses were rowing on the lake with a drag attempting to recover a lost net. Suddenly they were pulled into the middle of the lake at a great speed. A big white whirl appeared on the surface of the water and the boat suddenly stopped. One of the witnesses, a man named Emil, believed that there was a beast in the lake. [28]

SIMCOE LAKE, ONTARIO, CANADA

Length: 30 miles. Maximum Breadth: 18 miles. Area: 280 square miles. Reports of a lake monster here date back to the 1880s. The animal has been given the local name of 'Igopogo' or 'Simcoe Kelly'. A sighting in 1963 was of a thirty to seventy foot long, charcoal coloured creature with dorsal fins and a dog like face. [1] [41]

SINIAN LAKE, HUBEI, CHINA.

In the summer of 1981 there was an amazing sighting of a strange creature on this lake. Professor Chen Mok Chun and eight other scientists were setting up television cameras when they saw three

Loch Oich

Loch Shiel

huge toad-like creatures rear out of the pools and move towards them. They were reported to have greyish-white skins, mouths six feet across and eyes larger than rice bowls. One creature, then, opened its mouth and extended its tongue which it wrapped around the cameras on the tripods which it then reportedly engulfed! The two other let out screams and then all three disappeared back into the lake. [21]

SKORADALSVATN, ICELAND

In 1861 Baring-Gould was told by a friend of a sighting of a `Skrimsl` in the lake. Apparently a farmer had seen two humps and a head like that of a seal. Its length was said to be forty-six feet (a very precise estimate if I may say so), consisting of a six foot head and neck, a twenty two foot body and an eighteen foot tail. [42]

SNASA LAKE, NORWAY

There have been reports of a lake monster which is seen during calm weather. [1]

SOLIMOES RIVER, COLOMBIA

In 1967, Franz Hermann Schmidt and Captain Rudolph Pfleng were travelling up the river when they noticed the tracks of what appeared to be three separate creatures, and they also noticed that vegetation had been torn down. The following day they observed a creature near the shore. It had a barrel sized head on a ten foot long snake-like neck.. It had small eyes and a snout like that of a tapir. It was between eight and nine feet tall at the shoulder and possessed a pair of massive flippers equipped with claws. They fired several shots and apparently hit it although this seemed not to do the animal any harm. It then dived into the water revealing as it did so, a blunt tail with rough horny lumps. [7]

SSOMBI RIVER, AFRICA.

In 1913, a leader of a German expedition to The Cameroons was shown the path used (allegedly) by a `Mokele-Mbembe` to get to its food source. [26]

STAFFORD LAKE, CALIFORNIA, USA

Like Seattle Lake, this stretch of water has for decades been the source of lake monster rumours. The lake had been dammed in the 1950s and in April 1984 draining of the lake was started. In the august a six and a half foot long white sturgeon with an estimated age of between fifty and sixty years was found in a shallow area. It took twelve men to wrestle it out of the lake, and it was taken to Steinhart Aquarium in San Fransisco where, unfortunately, it died a week later. [21]

STORSJON LAKE, JAMTLAND, SWEDEN

This is the deepest lake in the country with an area of 176 square miles. For hundreds of years there have been reports of a creature living here and it has been given the local name of "Storsjoodjuree". Attempts were made in 1894 to capture the creature with a huge spring trap baited with a piglet. The creature is often described as having a ten foot long neck, a small head, very large eyes, and most unusually little head bone extensions often described as fins or ears. This has led Danish researcher Lars Thomas to suggest in a recent article, that some sightings of the "monster" are probably misidentifications of the local moose population swimming in the lake. [1] [7] [12] [19] [21] [32]

SUAINBAHL LOCH, ISLE OF LEWIS, SCOTLAND.

It is said that in 1856 crowds of people saw a monster in the loch. Its length was estimated as forty feet and it had the appearance of a giant eel. [1]

SUNDABARNS (mouth of the Ganges), INDIA

There have been reports here of monstrous animals up to twenty feet in length. Bernard Heuvelmans has suggested that they are possibly huge monitor lizards [37]

SUWANNEE RIVER (LOWER) FLORIDA, USA

This is the famous site of giant three toed tracks which appeared along the bank of the river in 1948. The fortean researcher and zoologist Ivan T Sanderson investigated the matter and believed that they were made by a giant penguin. On another occasion he believed that he had actually seen this beast which he described as being a dirty-yellow creature twenty feet long and eight feet wide and wallowing on the banks of the Suwannee Gables. In 1988 Tony Signori confessed to having hoaxed the tracks with the aid of cast iron `monster` feet. What (if anything), however, did Sanderson see? The sad truth is that we shall probably never know! [26] [49] [86] [87]

SWAN (CREEK), QUEENSLAND, AUSTRALIA.

In the 1860s or 1870s Mr T Hall observed a creature which the indigenous population of the area referred to as the `Mochel Mochel`. He heard a scream and saw an animal similar in shape to a sheep dog. Its head and whiskers resembled those of an otter and it`s colour was described as being that of a platypus. (The colour range of this species is between sepia brown to almost black).

It was passing from shallow water over a strip of dry land into deeper water. I believe that it could very well have been a seal. [29]

SYRACUSE MARSHES, SICILY

In December 1933 a reptilian monster was said to have been killed in the marshes by the local peasants. It was described as being scaly and eleven feet in length. Scientists, however, believe that it was an escaped python or boa. [41]

TAGAI LAKE, BRITISH COLUMBIA, CANADA

This shallow lake has a report of a lake monster seen in August 1976. According to the reports it first harassed a fisherman, but was later seen by three people who were standing on the shore. It was apparently ten feet in length and appeared to be moving just under the surface. Although these were the only two sightings ever reported it still earned the local name of `Tag`. [88]

TAHOE LAKE, NEVADA, USA

This lake monster, which has been given the local name of `Tessie` was brought to cryptozoological attention in 1986 when an anomalous finned object was allegedly filmed moving through the water leaving a wake of between twenty and twenty five feet. Unfortunately if the film ever existed (which seems doubtful), it has never been made available for scientific scrutiny. A popular explanation for the creature seems to be a giant fish such as a sturgeon. [21] [89] [90]

TAMANGO RIVER, ARGENTINA

In 1907 an explorer named Vaag, was exploring the river when allegedly he came across the remains of a huge animal and the tracks of another. [1]

TANA LAKE, ETHIOPIA, AFRICA

Area: 1390 square miles. Bernard Heuvelmans has suggested that an unidentified type of creature which is referred to be the local population as `AULI`, `AILA` or `IA-BAHRTED CHA`, and which reportedly lives in this lake could be an unknown species of Sirenian. [37]

TANGANYIKA LAKE, TANZANIA, AFRICA

This huge lake, four hundred miles long, twelve hundred square miles in area and with a maximum depth of 4,710 feet (the deepest in Africa), has a lake monster that the local fishermen have named

`PAMBA`. They say that it looks for canoes to swallow and that it is a huge fish-like monster with horns, the head of a bull. They also say that it spouts water like a whale. A more dinosaur-like creature was reported from this lake in 1928. A monster resembling an island and which left three clawed prints on the shore was observed from a distance by people on several ships. When they attempted to draw nearer, the creature sank beneath the waves. The prints were larger than those of an elephant and there was evidence that the beast possessed a thick tail. [7] [19]

TARLA LAKE, AUSTRALIA

George Holder, (who has already been mentioned regarding sightings in PAIKA lake), reported that some of his employees had seen a bunyip in this lake. [1]

TASEK BERA, PAHANG, MALAYSIA

The native name for the monster which has been reported from this part of south east Asia is `Ular Tedong`. A sighting which took palace in the early 1950s was by a senior officer in the Malaysian Police Force who whilst swimming, saw a fifteen foot long neck rise above some weeds about forty yards away from him. He could also see the contours of two silver grey curves on the surface of the water. The local population describe the creatures as having a snake-like head bearing two snail-like horns. They are said, when young, to have slate coloured scales which become golden as the creatures get older. They possess a long neck and a long tail and do not come on land. They are plant-eaters and occasionally emit a loud cry. Karl Shuker and Bernard Heuvelmans have speculated that the creature may be a surviving plesiosaur or sauriopod dinosaur [7] [17]

TASEK CHINI, PAHANG, MALAYSIA

A creature very similar to those reported from Tasek Bera has been reported from here. The local population claim that they are born on the top of a mountain called Gunong Chini, from whence they migrate using interlinked mountain streams. [7] [17]

TAY RIVER AND FIRTH, PERTHSHIRE, SCOTLAND

Length: 118 miles. In a letter to author Maurice Burton which was received sometime before the 1960s, a witness described a sighting of a horse-shaped head which reared up out of the water, turned from side to side and then plunged back down into the water. Late one night in September 1965 Maureen Ford was driving near the Firth of Forth when she saw a creature only yards from the river. It was a long grey shape with no legs and pointed ears.

The animal was observed for a second time two hours later after having (presumably) crossed the road. Robert Swankie saw the animal illuminated by the headlights of his car. Its head was over two feet in length and again appeared to have pointed ears. Its body was approximately twenty feet long and supposedly `humped` like a caterpillar, It moved very slowly and made a noise described as like someone pulling a heavy weight through the grass. [7] [36]

TAY LOCH, TAYSIDE, SCOTLAND.

Length: fifteen miles. River Tay flows through this lake, and a fourteenth century map of Scotland noted that in this loch there were three wonders - a moving island, fish without fins and waves without wind. [1] [12]

TELE LAKE, CONGO, AFRICA

This is said to be the haunt of what is undoubtedly Africa's most famous lake monster - The Mokele Mbembe. The favourite identity which has been hypothesised by researchers is that of a surviving Sauropod dinosaur, but it has also been suggested that the true identity of the beast is that of a giant monitor lizard. It has also been suggested that the Mokele Mbembe is a giant soft shelled turtle, or even an elephant crossing the lake by swimming with its trunk raised out of the water!

The lake first came to cryptozoological attention in the late 1970s and expeditions soon followed. First was the controversial Hermann Regustars Expedition, followed by that organised by Roy P.Mackal and Richard Greenwell. They discovered a story from the local population which told of one of these fearsome beasts which was killed with spears by a tribe of pygmies in the late 1950s. The victorious tribesmen then decided to eat the animal whereupon they all died.

Further expeditions included that of a Japanese film crew who in 1992 produced a controversial video of what they believed was a Mokele Mbembe crossing the lake. (This was, by the way, T.B.S - the same crew who also produced the controversial film of what purports to be the Lake Dakataua monster. Other notable expeditions were `Operation Congo` mounted by Bill Gibbons in 1986 and 1992. [7] [39] [40] [92] [93]

TIAN-CHAI (TIANCHTIANCHI) LAKE, JILIN, CHINA.

This lake, which is, by the way, inside the crater of a volcano, has produced over five hundred reports of a monster with the local name of `Changbai`, which date back over a hundred years. What makes this lake particularly interesting, however, is that the volcano erupted as recently s the beginning of the eighteenth century, so whatever it is that lurks within the lake must have arrived after that date, and even more extraordinarily, to an altitude of 6,400 feet! One of the most famous sightings was in August 1980 when a party of meteorological researchers saw a huge creature which was larger than a cow. It had an elongated neck more than three feet in length, a cow or dog shaped head, and a flat duck-like beak. One member of the group shot at it, apparently grazing the tip of its head, at which the creature dived beneath the surface of the water. [7] [21] [44] [72] [94]

TIBERIAS LAKE, TASMANIA, AUSTRALIA

There were sightings of a bunyip here in the nineteenth century. A sighting in 1852 was of a bulldog-headed creature about four or four and a half feet long with black shaggy fur. [25] [29]

TITICACA LAKE, PREU/BOLIVIA, SOUTH AMERICA

Length: 130 miles. Area: 3,200 square miles. Maximum Depth: 1,214 feet. This is the largest lake in South America and is home to reports of a twelve foot long creature resembling a seal or manatee [37]

TREIG LOCH, SCOTLAND

It was once said that the largest and fiercest of `water bulls` lived here. In 1933 Engineer, B.N. Pearch, who was in charge of the Hydro-Electric scheme claimed that some of the divers working on the project had reported seeing a monster deep in the lake. As a result, he claimed, some of the men had left his employment! [1] [18]

TURTLE LAKE SASKATCHEWAN< CANADA

Since 1924 there have been sightings here of a creature which locals believe is a large sturgeon. [10] [83]

TYIHUMBWE (UPPER) RIVER, CONGO, AFRICA.

In the 1930s two or three giant crocodiles were said to exist here. They were given the local name of `Lipata`. They were almost always hidden in the water and they very rarely emerged onto land, although they were said to be more active when it was raining. They were said to attack pigs, goats, cattle and even men which were swallowed in one mouthful.

Some of these beasts were reportedly killed and possessed a mouth larger than a normal crocodile - the throat was wider and the eyes were closer together! [40]

UBANGI RIVER, (MATABA TRIBUTARY)

There have been rumours here of a creature known as the `Nguma-Monene` by the local people. It is said to resemble a colossal snake at least 120 feet in length, but it has a serrated ridge along most of its body! It has a forked tongue and it walks on land with a low slung body! [7]

URABAHL (URUAL) LOCH, ISLE OF LEWIS, SCOTLAND.

In July 1961 three anglers fishing on the loch saw a strange creature forty five yards away from them. It had a hump about six feet away from what they believed was either a small head or a fin. It appeared three times and undulated, and the witnesses affirmed that it was much bigger than an otter. (1)

URI LAKE, SWITZERLAND

A monster is said both to have been seen and photographed here. (88)

URNER LAKE, SWITZERLAND

In August 1976 a long necked monster between twenty and twenty-five feet in length was seen surfacing three times by a crowd of about sixty people. Such a multi-witness sighting soon attracted media attention and the creature was soon dubbed `Urnie`. Unfortunately it emerged that the whole affair had been a hoax by a TV company using a sixty foot model. (It is interesting, however, to note that all the witnesses underestimated the size!) However, two divers then came forward, and stated that they had seen a similar creature in the lake the previous spring! [41]

URUGUAY RIVER, URUGUAY, SOUTH AMERICA

A large creature known as the `Minholao` (Giant earthworm) is said to live here. Bernard Heuvelmans has suggested that it may be a burrowing mammal, and Dr Karl Shuker has postulated a giant caecelian (a limbless amphibian) as an identity. [7] [37]

UTAH LAKE, UTAH, USA

Length: 23 miles. Breadth: eight miles. Area: 150 square miles. There were rumours in the nineteenth century of a lake monster here.

UTOPIA LAKE, NEW BRUNSWICK, CANADA

There have been reports since the early nineteenth century of a monster which has been given the local name of 'Old Ned'. In 1972 a serious attempt was made to capture it with the aid of nets and traps. The most recent sighting that I can find is from 1982 when the Hatt family saw a creature 'like a submarine' surfacing on the lake. It had spray on both sides, was approximately ten feet in length, and had a black rounded back like that of a whale. [41] [86]

VACCARES LAKE, CARMARGUE, FRANCE.

This great pool which lies between the mouths of the Grande Rhone and Petit Rhone was said by writer Joseph D'Arbano to harbour a monster, but as Ulrich Magin has reported, it is not known whether this account is based on truth or is wholly fictional. [41]

VAN LAKE, TURKEY.

The largest lake in Turkey has recently become the centre of lake monster attention because of the creature which has been seen in its waters. It has been reported by many witnesses, some of whom are of particularly high social standing - for example the sighting by Bestam Alkan, the provincial Deputy Governor, who in 1995 saw a black dinosaur-like creature which had triangular spikes on its back. More recently Udal Kozak, who has been investigating the mystery, took what appears to be video footage of the beast. At the time of writing this footage is being examined by experts at Cambridge University. [22]

VENACHAR LOCH, CENTRAL SCOTLAND

In 1800 the supposed monster of the lake was blamed for the death of several children drowned whilst crossing the lake. [5]

VICTORIA LAKE, KENYA/TANZANIA/UGANDA, AFRICA

Length: 250 miles. Breadth: 150 miles. Area: 26,828 square miles. This is the largest lake in Africa and the second largest in the world. There have been reports here of a lake monster known as the 'Lau' or 'Lukwata'. Sir Clement Hill was on a steam launch in 1902 when he saw the creature appear out of the water and attempt to seize a native who was sitting in the bows of his launch. He also saw it behind the boat. Its head was visible and it appeared to be rounded, fish-like and dark in colour. He was certain that it was not a crocodile. Both Bernard Heuvelmans and Karl Shuker have suggested that it is a presently unknown species of giant catfish. [5] [12] [21] [37]

VOROTA LAKE, SIBERIA, RUSSIA.

During the 1950s visiting geologists reported a monster in the lake. One report (from July 1953) was of a creature three hundred yards away. The foreparts of the creature, which the witnesses believed was the animal's head, had a breadth of two metres, and its eyes were set widely apart. The body was ten metres long and grey in colour. On its back was a kind of dorsal fin about half a metre high. the 'fin' was narrow and bent backwards. It undulated as it moved, before stopping a hundred yards from the shore, beating the water and plunging down out of sight [1] [12] [13]

WALKER LAKE, NEVADA, USA.

Length: 24 miles. Breadth: between two and six miles. In 1956 two witnesses were driving in their car when they saw a creature travelling at approximately 35 miles per hour. They saw it three times over a period of between ten and fifteen minutes. They said that it was between forty five and fifty feet long, its back being between four and five feet out of the water. [1] [10]

WALLOWA LAKE, OREGON, USA.

Before white settlers arrived in the 19th Century there were legends of a monster in the lake, and in 1982, Mr M Cranmer saw a creature about fifty feet in length with seven humps. [63]

WARREN COUNTY (ARTIFICIAL SWAMP ON FARM), NEW-JERSEY

In the early 1970s Ivan T Sanderson (yes, HIM again), and his wife reputedly observed a two foot portion of what appeared to be a large, pink, worm-like creature, in dense water-vegetation in a pond on his farm. [7]

WEMBU, (WEMBO) LAKE, TIBET.

Surface area: 310 square miles. Average Depth: 300 feet. In 1980 there were reports of a dinosaur-like creature seen in this lake which is indisputably rich in fish. It is said to have a long neck, a huge head, and a body as large as a house. The creature was blamed by locals for the disappearance of a yak from the shore, and also that of a farmer who vanished whilst rowing on the lake. [7] [21]

WHITE LAKE, PATAGONIA, ARGENTINA, SOUTH AMERICA.

In 1897 a Chilean farmer told of his sightings of a huge creature in the lake, It possessed along, reptile-like neck and disappeared at the slightest sound. [1] [7]

WHITE RIVER, ARKANSAS, USA.

Since the 1930s there have been reports of a monster on this river. It has been given the spectacularly imaginative nickname of `Whitey`. These were brought to the attention of the general public in 1971 when there were a number of sightings of a creature sometimes described as being as long as a car, a greyish colour, and sometimes possessing a long pointed bone which protrudes from its for head. There were even reports of three toed tracks being found on the shore. Cryptozoologist Roy P. Mackal has suggested that the creature is an aberrant and out of place Elephant Seal. [8] [83] [98] [99]

WINGECARRIBEE SWAMP, AUSTRALIA.

This is supposed to be the haunt of a fearsome bunyip! [20]

WINNEPEG LAKE, MANITOBA, CANADA

Length: 122 miles. Breadth: 17 miles. Area: 9,400 miles This huge lake is the alleged home of `Winnepego`. In 1935 Tom Spence and C.F.Ross saw a strange creature, dull grey in colour with a small flat head. There was a single horn protruding from the back of the creature`s head like a periscope. The rest of the creature`s body was likened to that of a dinosaur. [1] [49]

YOAN LAKE, (OUNIANGA SWAMPS), CHAD, AFRICA.

According to Dr Bernard Heuvelmans this is the probable home to an unknown species of Sirenian. [37]

ZEEGRZYNSKI LAKE, POLAND.

In 1982 this lake produced the report of a twenty foot monster which startled a swimmer. It had a slimy back with rabbit-like ears. Ulrich Magin and Karl Shuker both believe that it may have been a giant European Catfish or Wels - the size having been exaggerated. The `rabbit-like` ears could well, they believe, have been barbels. [21] [41]

ZWISCHENAHN LAKE, GERMANY.

In 1979 to Water Police Officers reported seeing a twelve foot long creature with a slimy back break the surface of the water. Further sightings were widely reported in the media. It has been suggested that as at Lake Zeegrzynski, Lake Zwischenahn could harbour a giant European catfish. [21] [41]

REFERENCES.

1. COSTELLO, P. In Search of Lake Monsters. (Garnstone Press, 1974)

2. MACKAL, Roy P. Monsters of Loch Ness (Futura, 1976)

3. McEwan, G. Mystery Animals of Britain and Ireland. (Robert Hale, 1986)

4. DASH, M. in FORTEAN TIMES (FT) # 52 (1989)

5. BORD, J and C. Writing in The Unknown (September 1985)

6. HEUVELMANS, B. On The Track of Unknown Animals - English edition (Hart-Davis, 1958)

7. SHUKER, K.P.N. In Search of Prehistoric Survivors (Blandford, 1995)

8. MACKAL, Roy P. Searching for Hidden Animals. (Doubleday, 1980)

9. Pursuit Vol 6, #2 (April 1973)

10. WELFARE, S. and FAIRLEY, J. Arthur C. Clarke's mysterious World (Collins, 1980)

11. The Unexplained (Orbis Partwork, Volume One)

12. MONTGOMERY-CAMPBELL, E and SOLOMON, D. The Search for Morag. (Garden City Press, 1972)

13. Animals & Men (A&M) #6

14. FT #82

15. HOLIDAY, F.W. The Dragon and The Disc (Sidgewick and Jackson, 1973)

16. The Unexplained Volume 16 (Orbis Partwork)

17. In Britain #4 (April 1973)

18. HOLIDAY, F.W. The Great Orm of Loch Ness. (Faber and Faber, 1968)

19. MUERGER, M. lake Monster Traditions. (Fortean Tomes, 1988)

20. BORD, J and C. modern mysteries of Britain (Grafton Books, 1987)

21. SHUKER K.P.N in Fate Vol 43, #9 (1990)

22. A&M #9

24. FT #84

25. SHUKER, Dr K.P.N. The Unexplained (Carleton, 1996)

26. SANDERSON, I.T. More Things (Pyramid)

27. Pursuit, Vol. 144 #3 (Autumn 1981)

28. BORD, J & C. Alien Animals (Granada, 1980)

29. HEALEY, T and CROPPER, P. Out of the Shadows (Ironbank, 1994)

30. BENEDICT, W.R in Strange Magazine #5 (1989)

31. Pursuit Vol 12 #2 (Spring 1979)

32. A&M # 10.

33. A&M # 13

34. FT #100 (1997)

35. BORD J and C. The Unknown (1985)

36. BURTON, M. The Elusive Monster. (Hart-Davis, 1961)

37. HEUVELMANS, B. in Cryptozoology #5 (1986)

38. ZARZYNSKI, J.W. Champ - Beyond the Legend (M-Z Information, 1988)

39. WELFARE, S. and FAIRLEY, J. Arthur C. CLarke's Chronicles of the Strange and Mysterious (Collins, 1987)

40. MACKAL, R.P. A Living Dinosaur (Brill, 1987)

41. FT # 46

42. EBERHARDT, G.M. Monsters (Garland, 1983)

43. FT # 78 Dec/Jan 1995

44. A&M #5

45. SHUKER Dr K.P.N writing in Strange Magazine # 15 (1995)

46. SHUKER Dr K.P.N. pers comm 31.1.95, 2.8.95.

47. NAISH, D.W. in CFZ Yearbook 1996

48. WHYTE, C. More than a Legend. (Hamish Hamilton, 1957)

49. DINSDALE, T. The Leviathans (Routledge, Jegan and Paul 1966)

50. News Herald 30.9.1990

51. ISC Newsletter (ISC) Vol 5 #1 Spring 1986.

52. Strange Magazine #8

55. SHUKER. Dr K.P.N. (Collins 1993)

56. The Sun 15.9.1997

57. The Star 15.9.97

58. BEAVAN S (Sea Life Centre, Portsmouth), Pers Comm 20.10.87

59. SHUKER Dr K.P.N Pers Comm, 6.1.90

60. FT # 61 (February 1992)

61. Nessletter (NL) #107 (November 1991

62. Alaska Magazine, January 1988

63. ISC Vol 6 #2. 1987

64. ISC Vol 5 #4. 1996

65. SHUKER, Dr K.P.N. Mystery Cats of the World (Hale, 1989)

66. FT #92 (November 1996)

67. ISC Vol 1 #1 (1992)

68. A&M #11

69. A&M #7

70. NL #112

71. FT #67 Feb/Mar 1993

72. Pursuit Vol 3 #1 (Jan 1970)

73. ISC Vol 6 #2 (1987)

74. BOISVERT, J. The Sea-Serpent of Lake Memphre-Magog (1983)

75. Pursuit Vol 15 #2 (Spring 1982)

76. FT #102 September 1997

77. FT #77 Oct/Nov 1994

78. DINSDALE, T. TheLoch Ness Monster (Rotledge, Kegan and Paul 1982)

79. ISC Vol 5 #2 (1986)

80. ISC Vol 6 #1 (1987)

81 You Magazine Supplement, Mail on Sunday (5.11.89)

82. FT#57 (Spring 1991)

83. ISC Vol 1 #2 (1982)

84. Cryptozoology Vol 3 (1984)

85. ISC Vol 4 #4 (1985)

86. FT #66 Dec-Jan 1993

87. ISC Vol 7 #4 (Winter 1988)

88. Pursuit Vol 11 #4 (Autumn 1978)

89. NL #70 (June 1985)

90. NL # 108 (Dec 1991)

91. Strange Magazine No 6

92. ISC Vol 2 #4 *1983)
93. FT #86 (May 1996)
94. A&M #12
95. BAUER, H.H. The Enigma of Loch Ness (University of Illinois Press, 1988)
96. FT #85 (Feb-Mar 1996)
97. Uri Geller's Encounters (September 1997)
98. ISC Vol 1 #3 (1982)
99. ISC Vol 2 #1 (1983)

Nyaminyami
The River God of the Zambesi

by

Chris Moiser

The Zambezi is one of the great rivers of Africa. It rises in the North West of Zambia and runs 1,650 miles to the sea. Its course runs South through Zambia, then it forms the Zambia/Zimbabwe border, before passing through Mozambique and then running into the Indian Ocean via a delta near Chinde. It has along its length two wonders of the world, one natural, the Victoria Falls, and the other man-made, the Kariba Dam, which in turn created Lake Kariba.

Victoria Falls, or Mosi-oa-Tunya (The smoke that thunders) was named after Queen Victoria by the Scottish explorer Livingstone, who was the first European to see them in 1855. At this point the river is just over a mile wide and drops four hundred feet to run through a gorge that is just one hundred feet wide.

Two hundred and forty miles downstream from the falls is the Kariba Dam. It was built between 1955 and 1959 across the Kariba Gorge to supply hydro-electric power to what was then Northern Rhodesia and Southern Rhodesia, and is now Zambia and Zimbabwe respectively. In addition it was to create lake Kariba in the Gwembe Valley, establishing a commercial fishery, and what is now a major tourist attraction and holiday resort.

As the Gwembe valley flooded so the Lake formed. Although the Dam was finished in 1959, it was September 1963 before the lake finally reached its maximum height, and final size of 180 miles long, and forty miles wide at the widest part. It was known from the beginning that in flooding the valley a large quantity of wildlife, and somewhere in the region of 57,000 local people would be displaced.

The rescue of the animals is well documented, and there is a monument to operation Noah at Kariba Heights listing the nearly 5,000 animals that were rescued from the rising waters by the Southern Rhodesian Wildlife Department as it then was. In Northern Rhodesia the game was less dense and there only 2,000 or so animals were rescued. The plight of the people in the Kariba Valley is less well documented.

When it was decided in 1955 to go ahead with the scheme to build the dam and flood the valley the population that would be displaced were almost entirely members of the Batonka tribe, with possibly only three or four Europeans in the area, there being no farms there. The Batonka tribal name has changed slightly over the years, from the almost "Carry On ...." name of Batonka, through Batonga to Tonga.

The Tonga people had lived in the Gwembe valley for over 400 years. Their oral tradition contains no history of migration, and the oldest traces of them are in the Zambian Southern Province, dating back to the 11th century, so perhaps they have always lived close to the Zambezi. The inhospitable nature of the land in which they lived, together with a reputation for ferociousness meant that they were rarely bothered by other tribes.

Some of the African stories of head-hunting come from this area, and at least one recent account of the building of the Kariba Dam states that a chiefs favour could be gained by bringing him the head of travellers from within his area. The skulls would be mounted in his Kraal, the greater the number of skulls the greater the chiefs distinction. (Clements 1959).

These people worshipped a river god - Nyaminyami - who was not just a mythical river monster like a

177

dragon in a folk tale but a far deeper and subtler being. The Tonga have an awareness of a supreme and invisible being who is a ruler of the universe and of the living and the dead. Although they never claim to communicate directly with Nyaminyami, as the river god of the Zambezi he is no picturesque fiction, but a personification of supernatural power.

In nearly all the folk stories that relate to Kariba there is one thing in common about Nyaminyami. He is said to have his "headquarters" in a cave in a rock that thrust out of the water at the entrance to the Kariba gorge close to the dam site, to some even the rock is sacred. With the construction of the dam, and the flooding of the valley this rock is now under almost one hundred feet of water.

Despite a reasonable compensation deal involving a new allocation of land and cash payments a resistance to being moved arose amongst the Tonga, and was fed by the nationalist movement from the North. The agitators suggested that provided his people did not desert him, Nyaminyami would protect them. He would cause the waters to boil, and destroy the white man's puny structure at Kariba; alternatively he would confer on the faithful the ability to live underwater when the floods came. Magic papers were sold to the Tonga to give them protection. Interestingly the cost of magic rose with the rank of the purchaser. A child had to pay 1/9d(9p), a woman 2/6d(12½p), a man 3/6d(17½p), a village headman 10/-(50p) and a chief £3.00. Most bought them. The papers sold to the chiefs were not any sort of spell or prayer to Nyaminyami but turned out to be copies of a petition that had originated from a government minister in Northern Rhodesia asking the Queen to abandon the Kariba scheme.

Nature worked to strengthen the idea that Nyaminyami was protecting his people though, when in 1957, and again in 1958 there were totally unprecedented floods which caused serious harm to the coffer dams at Kariba. Despite the damage the Italian construction company were able to continue building the dam almost on schedule. Although there was continued feeling against moving most of the Tonga villages in the Southern Rhodesia zone were moved with little trouble. Usually there was a ceremony on the eve of departure when the river spirits were told the reason for their subjects departure.

By the end of 1958 most of the Tonga people in the Southern Rhodesian sector of the valley had been moved to their new homes on high ground, above the projected lake shoreline. In Northern Rhodesia though the authorities continued to talk and hardly anybody was moved. As the talks continued so the Tonga became more suspicious. Bush clearance was taking place, and it seemed an obvious start for the development of colonial farms. In fact the clearance was so that when the lake had formed trawlers would be able to operate in the area without snagging their nets. Further confirmation of colonial trickery was seen though when an airstrip was built and houses appeared. The airstrip was built as a short term measure for the contractors clearing the area and the houses were prefabricated, again for temporary contractors. The situation rapidly became critical and boiled over at the village of Chisamu when the police, trying to evacuate the village, came under attack. Eight Tongas were killed and fourteen wounded by police officers firing in self defence.

Sir Arthur Benson, the Governor of Northern Rhodesia, blamed the African Congress for continuing to spread the rumours that the Kariba Dam enterprise was simply a conspiracy to deprive Africans of their land After the deaths though the situation calmed down and the rest of the tribespeople were moved out still reluctantly, but peacefully.

It might be thought that the completion of the moving out of the Tonga people would have been the start of the decline in the belief in Nyaminyami. In fact if anything he is now, almost forty years later known by more people than ever before.

The Kariba development has created a series of tourist resorts, with a large number of houseboats on the lake and a number of hotels close to the dam. The fishery has developed and is staffed largely by people of the Tonga tribe, who are now fishing commercially and making a reasonable profit. Other members of the tribe are working in the hotels and working on local farms. Souvenir production for the tourist trade is also a source of employment for many local people. Elaborate wooden carved walking sticks are one of the most popular items, they are made in the image of Nyaminyami and sold to both visitors and locals.

Nyaminyami is said to still live in the lake. Tourists and locals occasionally claim to see him. Some say he is like a whirlwind, but the majority say that he is dragon-like with a snake's torso and a fish's head. He is now said to be a brooding god who swims the lake waiting for an opportunity to destroy the dam. Now though he is not going to destroy the dam to finally return the Gwembe valley to the Tonga, but to be re-united with his wife. She was said to be downstream when the dam was constructed and has been separated from him ever since.

References

Clements, Frank. 1959 Kariba: The struggle with the River God. Methuen. London.

Lagus, Charles. 1959 Operation Noah. William Kimber. London.

Martin, David. 1996 Kariba - Nyaminyami's Kingdom. African Publishing Group. Harare.

A carved statue of Nyaminyami outside the Kariba visitors centre, with the dam in the background

AFRICAN STORIES

BY ROY KERRIDGE

Not long ago, a West End art gallery was the venue of a glittering scene - a Surtees Society party to launch their republication of Joel Chandler Harris's "Uncle Remus" stories. As a boy I loved these stories of Brer Fox and Brer Rabbit. The phonetically-spelled, homely southern dialect of the ex-slave Uncle Remus appealed to the depth in me.

It was a sobering thought, amid the sparkle of champagne, that I and the various other old buffers on display would almosxt certainly be the last generation to read the original Uncle Remus stories for pleasure. Once regarded as a children's book, part of every nursery, the dialect-stories of "Uncle Remus" will henceforth be banished to the turgid archives of academe. The dialect has become an insuperable barrier to modern readers. Educational advance (or lack of same) has made "Uncle Remus" unreadable for the under fifties. This is all the more odd when you reflect that the pseudo-Jamaican dialect poems of Linton Kwesi-Johnson and others were required reading for unhappy primary school children as little as eight years ago.

"Uncle Remus", for those not in the know, is a composite figure, a kindly old man whom Harris depicts as telling African-derived folk tales to a little white boy for more than a thousand and one nights. In reality the stories had been painstakingly collected by Harris from Negroes all over the state of Georgia, before and after the American Civil War. The first of many "Uncle Remus" volumes appeared in 1881, to great acclaim. This is the book reproduced by the Surtees Society, who normally specialise in fascimile editions of Victorian and Edwardian sporting novels. Brer Rabbit might well be considered a fox hunter, since he usually succeeds in turning the tables on his enemy Brer Fox.

Nearly every plot in an Uncle Remus story appears again and again in the comic animal trickster stories of Africa. I am lucky enough to have a pen-friend, Louis Baba of Ghana. A Cabinet Minister once came undercriticism at Westminster for speaking flippantly of `bongo bongo land' as if this must be an insultingly named imaginary African country. Nevertheless, Louis is a member of the Bongo tribe from Bongo-land, in the north of Ghana. In his job as snake, bird and animal catcher for the British pet-trade, he has travelled every kind of wild country within the vast imperial boundaries of Ghana. Wherever he goes, he collects local stories, legends and songs, translates them into English and sends them to me.

According to Louis Baba, animal fables must be told only after dark. In the north of Ghana, the trickster hero of the stories is the hare, in the south it is the Spider-Man, Anancy. By this information the provenance of slaves transported to the New World can be discovered. To this day Anancy stories are told in the Carribean, and Hare stories in the American South. For Americans the words "hare" and "rabbit" are interchangeable. The African Hare has become Brer Rabbit. Similarly the blues songs of the American South can be traced to the music played by "griots" (troubadors) of the northern West African savannah country, home of the African hare. Such hares thrive on grassland, and cannot live in mangrove swamps, rain forests and other spidery places.

Uncle Remus's tales of Brer (Brother) rabbit, Brer Possum, Brer Bear, Coon, Buzzard and other American species came from many widely separated West African sources. Only when English had become a common language for an entire generation of American slaves could the stories have taken the form in which Chandler Harris found them, with African animals transposed to the New World counterparts. Although most stories of the New World African diaspora must have come from West Africa, the same trickkster tales are told all over Africa, south ofthe Sahara. South of the Kalahari, the sage hero is not a hare, nor a spider but a jackal. The plots of the stories remain the same.

When my mother crossed the Kalahari by landrover in 1990, her guide was an elderly man called George Lehkaukau. George could speak Tswana, English and a smattering of Bushman. In his spare moments he trapped jackals and made elaborate and beautifuol robes from their fur. Once only, to entertain my mother, he told jackal stories. Told as they should be told, these stories come alive, complete with songs, realistic animal noises and minute observations on real-life animal character and behaviour. Louis Baba has sent me tapes of Bongo songs-from- stories. Heard without the stories, these meant little to me. George's stories and those of Uncle Remus are clearly blood brothers. Remus stories also come spiced with songs, animal noises and naturalist observations.

In chapter six of "Uncle Remus" we find that "Mr. Rabbit grossly deceives Mr. Fox". Boasting to his friends, the Rabbit speaks dismissively of the Fox as "my daddy's riding horse". Hearing of this later, the enraged Fox, gnashing his teeth, flew to the Rabbit's house with the intention of making him publicly eat his words. After which, of course, the Fox would privately eat the Rabbit.

Things worked out very differently, because the Rabbit refused to be enticed out of doors, claiming to be ill. Finally he agreed to accompany the Fox if he could ride on his back. Still feigning friendship, the Fox agreed, and allowed himself to be saddled and bridled up for the journey.

"When we're on the road, I'll shake off the Rabbit and catch him in my jaws" he reasoned to himself.

It was not to be for the Rabbit came equipped with spurs, and so he gained mastery over the Fox. He bucked and hallooed his way past his friends`s house on the Fox`s back in fine rodeo style. How the Rabbit escaped, having impressed his friends, is another story.

Deep in the Kalahari, George told the same story with the Jackal as the trick rider and the leopard as the duped "horse". George, himself, owned several donkeys, each proudly arrayed in beautiful ornamental bridles and beaded horse tackle made only in the huts of the Kalahari. In pidgin-English, the Leopard is known as the "Tiger" all over Africa, and so he was called by George. "Brer Tiger" is a popular figure in Carribean folklore, whilst Savannah-loving Kong Lion roars his way through stories from the American South. Such big cats have no Southern or Caribbean counterparts. Jaguars in Guyana are also known as "Tigers".

Brer Anancy, in Jamaica, in one story was set the task of capturing a snake and a swarm of bees. He persuaded these creatures to capture themselves in an ingenious fashion. Taking a stick, he wondered aloud, if it were, or were not, longer than Brother Snake. Proudly, the snake (who had overheard) claimed to be longer than the stick and lay down beside it.

"No, no, Brer Snake, you are cheating! You are sliding forward to seem longer than this stick!", cried Anancy.

Indignantly, the serpent denied this charge and agreed to be tied to the stick to prove his length with no cheating. So that was the snake captured! Using the same technique, Anancy took an empty bottle to a beehive and stood musing over it, "a swarm can`t go in! No it can`t!"

Helpfully, the bees sought to prove that they COULD fly into the bottle. Whereupon Anancy produced a cork....

George told exactly the same story, with the Jackal taking the place of Anancy. In America, Brer Rabbit was set similar tasks by his fearsome relative the Witch Rabbit, Auntie Mammy Bammy Big Money.

Obtaining a copy of "Uncle Remus" from the Surtees Society people, I posted it post-haste to Louis Baba in Ghana. Louis Baba was NOT impressed by Uncle Remus. He not only took Remus seriously as an individual story teller, but made the mistake at first of supposing the stories to be newly arrived from Africa. He credited Uncle Remus with the task of CONSCIOUSLY Americanising the stories, a process that actually took hundreds of story-tellers several generations to achieve. So, in one letter he both flattered Uncle Remus as a translator, and traduced him as a story teller. Many of the "Just-So-Story" explanations about African animals made no sense in America and had been abandoned by story tellers. Sly anti-authority slave humour obviously touched no chord with Baba, a free man of a free people.

The three stories that most interested Mr Baba were "The End of Mr Bear", "Mr Wolf makes a Failure", and "Miss Cow falls a Victim to Mr Rabbit". As told by Uncle Remus, the tale of Mr Bear does seem rather pointless. Lured to a hollow bee tree by Brother (or Brer) Rabbit, the Bear climbs up and puts his head into a hole. meanwhile the Rabbit pokes a stick up into a tree through another hole, further down, and stirs up the bees. They sting the Bear's face so much that it swells and he can't withdraw it from the hole. laughing cruelly, the Rabbit leaves him to his fate.

"Mr Wolf makes a failure" by making a pact with the Fox, in order to catch the Rabbit. Brer Fox lies on his bed and pretends to be dead, while the Wolf hurries to the Rabbit's house in the guise of a mourner. Summoned to the Fox's bedside, the Rabbit stays cautiously at a distance, remarking that dead folk always raise a back foot and shout "Wahoo!!" When the Fox shakes a paw and shouts "Wahoo!!" the Rabbit runs away laughing.

"Miss Cow" is one of my favourite Brer rabbit stories. She has never allowed herself to be milked by any of the local wild animals. Brer Rabbit, after much polite conversation, in the African "Convey My Greetings" style, persuades the Cow to but a persimmon tree to shake down some fruit. But the 'simmons are green, and are not ready to fall. So the Cow, eager to oblige, charges head-on at the tree, and her horns stick into the wood. Fetching his entire family, all with pails, Brer Rabbit milks her dry. It is midnight before Miss Cow manages to twist her horns out of the tree. She eats, drinks, and rests, and then cunningly manages to fit her horns back into the tree to await the Rabbit's return. When he arrives, she chases him furiously. He escapes by jumping into some mud with only his eyes showing.

"Howdy, Brer Big Eyes. Is you seed Brer Rabbit go by?" Miss Cow enquires...

Pointing onwards, the Rabbit watches the Cow run out of sight, then shakes off the mud and gos home laughing once more. We shall now see what Louis Baba makes of all this.

"..... Now to "Uncle Remus". The stories are most poorly told! There are no narratives, as compared with their native form. Perhaps if I didn't already know the stories in their native nature I might not be so disappointed. There stand a mere nine stories in the whole text. One Bongo story, six Northern stories, one Akan and a yarn we told at school - distorted here but not cleverly done.

I see them plain. Successively Wolf is cast for Hyena. When Wolf had to hunt for honey, the narrator obviously bearing worldly facts in mind, produced Bear for that task. But whereas in the Bongo account, bees stung Hyena, giving him an ugly mouth and ending his lust for honey, Uncle Remus's bear keeps his snout plus his appetite for honey.

In the Remus series, the stories are cast in shreds, pieces scattered between the covers. Not a single story, however peiced together, produces the meaningful ending of a Bongo story. Though I can understand, it is still a pity that characters or objects cast in our native narratives have to be changed to the nearest substitutes in the foreign country. Inevitably this leaves gaps for questions. On the other hand it does little harm: the entire text makes it clear that the man Uncle Remus was not a native of anywhere the stories originate. he might have got thm down as fourth hand, when they had already taken beatings in recycling.

In the story "Mr Wolf Makes a Failure", the inadequacy isn't very serious. Actually, when Hyena and Hunting Dog hatched the plot to devour Hare, the conspirators in their greed didn't think about the natural illness preceeding death. So when Hare deigned to go and see he was already precautious..."

(Louis Baba repeats Uncle Remus almost word for word, except that the Wolf is a Hunting Dog and the Fox a Hyena. Instead of a back leg, a front leg is raised pointing skywards to a heavenly destination. Hare runs home laughing as usual!)

".... That story cautious about tripartite treaties.

"Miss Cow falls a victim to Mr Rabbit" is the story I want to give in its native form"

continues Louis Baba..

"....but I hope that you let me know in your next letter what is a 'simmon tree. Does it resemble a Baobab tree? The Bongo shepherds were very wise when they picked a Baobab tree for this story. No other tree would have fitted so perfectly for the purpose.

Hare had obtained millet, he had obtained honey too. He dipped the millet in the honey, crammed the lot in his cheeks, and held a fistful more, as he set off towards Mrs Buffalo's lair.

Face to face with the dangerous woman, hare feigned surprise. Questioned, he said that he had strayed in a terrified flight from Hyena, who was seeking to bully him into revealing the source of the delicious nuts he had discovered. With that, Hare offered Mrs Buffalo some honey-covered millet.

"Delicious," the woman cried. "You must show me the tree that bears such lovely nuts."

Playing reluctant, Hare led her to a baobab tree. Then he took paces far back, rushed at the tree with terrific speed, leaping up high as he made contact with the tree trunk. In reality he only let his legs and ears hit the tree lightly. In the blur of a somersault, he sprayed honey coated millet out from his cheeks. He then picked them up. Mrs Buffalo then ordered Hare away from the scene. Hare didn't really go as far as he made the woman believe. He sneaked into a thicket in which he had already hidden a calabash pot.

Mrs Buffalo determined to rock the tree real powerfully - not merely to collect a few nuts, but to harvest them all. She disappeared to a far distace, and then charged the tree strongly. Bang! With her full strength behind her, she drove her horns deep into the trunk. So, she was held fast incapacitated and helpless.

Grinning, Hare then appeared from the thicket clutching his pot. As he sat beneath the woman, pulling at her teats, he sang along "Neloom Koom Pi!" (Bongo song about milking follows). Off he went home with the fresh milk. The next day at sun-tilt he went and drew another pot full.

For three days running he milked Mrs Buffalo whose stomach shrunk in while her ribs rayed out. It was safer to milk her to death than to scheme a plan to free her and contend with her everlasting vendetta.

In the early hours of the fourth day, the strongest army in the world came marching paston another of their campaigns of conquest. They were led by their General Feld Marshall whose blackened whiskers bounced wonderfully upon a face with glowing eyes. This division upon division of army was brought up in the rear by an equally disciplined but admirable Rear Admiral of the Fleet. This is the only army in the world that has never known defeat, nor suffered a single casualty in any of their daily attacks.

Which other army goes to war with all their pregnant women, all their babies and ageless old men to savour the joys of victory? The one army that would march across water without building a bridge! They came marching past.

Mrs Buffalo cried out to them for help. general Field Marshall Termite sent a dispatch rider to go and find out if this was a mercenary job. In the negotiation it seemed that there wouldn't be a payment from Mrs Buffalo. With what could she pay termites? But the old men made a proposal.

A single battalion then set to work, as the main column waited. Within minutes the termites had bored a perfectly neat socket around the horns freeing the woman at last. It was still morning, so

Mrs Buffalo staggered to the stream and drew in water. She ate voraciously and then drank again. She ate all the way back to the baobab tree by noon. She replaced her horns into the loose socket holes and stood waiting.

Here comes Hare at Sun Tilt. He detected the improved belly-shrink. All the same he slapped her flanks, squatted under her as usual and pulled those teats. Suddenly a hoof shot up, missing Hare's head by a breadth and sending the pot sailing through the air. Hare leaped, and took off at full speed. The woman charged after, propelled by sheer fury!

No doubt speed was on hare's side, but durability stayed with the vengeful woman. hare executed careful zigzags before leaping upwards. Landing on his back, he rolled into a ball. Ears beneath, he popped his eyes out. Dramatically he had transformed into a strange creature.

When Mrs Buffalo ccame upon this thing, she enquired urgently. "Mr All Eyes did you see that Hare pass that way?"

"There" pointed the transformed creature. the charging woman rumbled away leaving Hare to skip over the quagmire to the place where he is found today eating Okra beyond the search of Mrs Buffalo.

Today, too, you will see Buffalo rubbing against the termite mound. That is the offertory she performs evry time a buffalo passes by the termite's home. Get closer and you will see hairs left on the mound indicating that Buffalo made a sacrifice!

I know this is true, because I was there and I saw it happen!"

His story over, Uncle Baba - I mean Louis Remus - concluded with a few remarks on the Baobab tree.

"Now shall we see why there couldn't be an alternative to the Baobab tree in the story? The baobab is the widest tree in this land. Instead of wood it is a mass of fibre saturated by water. Should there be a hundred year drought only the baobab would not go wanting. Nothing can fell it. If a great branch breaks off, instead of drying out it rots like a human corpse. So it is related to Man rather than tree, just as the hyrax is related to the Elephant not the rodent.

Beetle larvae are weaned in decaying bodies of baobab. It is the most worshipped tree. Every baobab tree has her own special name. After the baobab comes the silk cotton tree.

The best tree to climb when in danger from a leopard is the baobab. If the leopard tries climbing, its claws sink into the tree's flesh, requiring effort to pull them out. (A woman is said to have plucked a fruit and then corageously hammered a climbing leopard's head until she killed it). This tree has flesh for bark. However, you must climb a different tree in cases of elephant pursuit.

I think it is now well to say, yours sincerely,

Louis Baba".

So ended the most interesting commentary on "Uncle Remus" that I have read for a long time.

Meanwhile back in the Kalahari Desert, Uncle George Lekaukau has one more story, about the jackal, the hyena and the leopard. When telling the story in English, George calls the leopard a tiger. Some South Africans call the hyena a wolf, doing an Uncle Remus in reverse. Leopard and hyena are good enough for me.

Hyena is big and strong, and jackal is small but cunning. Hyena hated jackal for playing tricks on him. Now, when a white man hates somebody, he tells that person so to his face and then shoots him. But when one of my people hates a man he talks VERY VERY sweetly to that man, VERY VERY sweetly, then in secret he kills that person by witchcraft. Hyena did not try witchcraft. He talked sweetly to Jackal and persuaded him that he knew a cave far off where there was a great store of honey.

Jackal followed Hyena and at last they reached a cave in the side of a rocky kloof (escarpment).

"Go in, the honey is in the back of the cave", Hyena insisted.

Jackal went inside, and Hyena rolled a great boulder over the mouth of the cave, leaving only a small crack through which he taunted Jackal. The crack was too small for Jackal to escape.

"Ha, Jackal, you have played your last trick on me. This cave belongs to Mrs Leopard, and when she returns she will kill you! Goodbye forever freind Jackal!"

Jackal ran up and down inside the cave. He could not move the boulder. He called Hyena, but Hyena had returned to his home. In the rear of the cave, Jackal heard a noise. There he found five new born leopard cubs.

"I would like to eat one of you fine fellows" Jackal mused, "but first I had better await your mother".

Sure enough, here comes Mrs Leopard, growling and coughing with anger at the sight of the boulder.

"Who has draped this stone over my doorway?" she exclaimed.

"If you please. Mrs Leopard, it is I, the jackal" answered Jackal. "Hyena was trying to eat your cubs, so I ran into the cave and dragged the rock after me to shut Hyena outside. All five of your cubs are safe. Look, and you will see Hyena`s footprint! Now, can you move the rock? I wish to go home".

"Hmm Hmmm" said Mrs Leopard, swishing her tail. "Yes, here is Hyena`s tack. Well done, Jackal! You have saved my cubs! But I will not allow you to leave, you must stay and guard my cubs while I go hunting! First I will give them suck!".

So, Jackal pushed the five tiny cubs through the crack. The mother fed them and pushed them back inside again. Then she went hunting, she was gone a long time. Jackal felt hungry.

"One little cub will make no difference" he said to himself.

He ate one of the cubs. Just then, Mrs Leopard returned.

"Hand me out my cubs for another feed" she commanded.

"Mrs Leopard, I think that I should hand each one out to you and take it back, one at a time", said Jackal. "That way they won`t fight for the best place, and all can have a fair share of milk".

"Good idea, Jackal!" said Mrs Leopard.

So Jackal pushed out one cub at a time, took it back and then pushed out the next one. Soon all four cubs had been fed.

"Now for the fifth cub," said Mrs Leopard.

Here was the difficulty! Jackal had eaten the fifth cub. So he took the first cub and pushed it out again. Mrs Leopard couldn`t tell the difference. She went away again, having pushed the cub back through the crack to Jackal. Jackal licked his lips as he contemplated the fat little cub - the one who had enjoyed two meals! So he ate that cub!

Every time the mother leopard returned, Jackal tried the same trick and ate another cub! By the following day there was only one cub left.

"How are my cubs today?" asked Mrs Leopard..

"All very well apart from the fifth cub," said the Jackal. "He looks a little sickly to me".

"Hand out the first cub then take it back" ordered Mrs Leopard.

So, cunning Jackal handed out the cub. It took suck and then was pushed back inside. At once Jackal pushed it out through the crack saying "Here is the second cub!"

In and out the cub went. Soon it had been fed four times. "Here is the fifth cub - this is the one who is sick!" cried Jackal.

"Yes, Jackal, you are right! The cub is crying! It refuses suck! What is wrong?"

Jackal knew that after four meals it was so full it could suck no more, and its stomch pained.

"Mrs Leopard! The cave has no air! That is why the cub is ill. Push away the boulder from the mouth of the cave!" Jackal cried.

In alarm, Mrs Leopard agreed. She pushed the rock UPWARD until it stood safely balanced on a ledge of stone above the cave entrance. Just then, she saw a buck in the distance and set off in stealthy pursuit, without examining her cubs.

"Now, plump one, you are mine!"

Greedily he devoured the cub. At that moment Mrs Leopard returned without her prey. Quick as a flash Jackal leapt up and held his front paws against the overhanging rock.

"Mrs Leopard! This rock is about to fall and crush your cave in an avalanche!" he cried. "Hold it quickly, while I rush out and fetch a stick to prop it up with."

Mrs Leopard ran forwaard and stood with her front paws pressed against the rock. She looked up and saw clouds gently sailing through the sky over the top of the cliff, giving the illusion that the whole cliff was swaying over, about to crash.

"Hurry, hurry," she roared, as Jackal made a nimble getaway.

Later, when Hyena asaw Jackal he could hardly believe his eyes.

"All night I fought the mother leopard until I finally beat her!" Jackal told his enemy. "Now I have gained such fighting skill I can beat YOU, Hyena!"

Hyena fled in terror, saying "Jackal is too much for me!"

As for Mrs Leopard, she held on to the rock and waited and waited for Jackal to come back. As far as I know she is still waiting!

So ended George's story. As I have heard it third hand without the songs and the animal noises, I cannot atempt the full flavour of an African story. Probably there ought to be a moral at the end. Ever education-conscious, Africans like their stories to have messages and meanings beyond mere entertainment.

Incidentally some witchcraft CAN kill! That type of witchcraft is known and used in England, under the name of Poison. Not all magic is good for you but the magic of African stories is one tonic I could take again and again!

NOTE: "Uncle Remus" by Joel Chandler Harris, can be obtained from the Surtees Society, Tacker's Cottage, Horn Street, Nunney, Nr Frome, Somerset, UK.

Some Strange Snake Stories
by Richard Muirhead

Snake Colouration.

According to scientific orthodoxy, which in the main is accepted by the layman, we have three species of native British snakes with none in Ireland. Furthermore in theory their colours are fairly consistent and predictable, with no major surprises to be expected. But the situation in reality is far more complex. In fact an entirely different species of snake from the adder may exist, the small red adder.

To complicate matters further, within the species known as the adder, there is a far greater range of coloration than the usually mentioned grey-brown hues. For reasons best known to more experienced academics, these interesting colour variations are not as well known as they should be. G.R Leighton, in his Life History of British Serpents, 1901, proved to his own satisfaction that the small red adder not only existed, but that it wasn't simply a colour phase of the juvenile female "common" adder but a species in its own right [1]. Even leaving to one side for one moment the many differing hues of the adder/viper there are interesting facts such as melanism and albinism in grass snakes and melanism in adders. At present I have no examples of unusual colours in the smooth snake, (owing to lack of research into this animal.) This does not necessarily mean there are no colour variations.

THE ADDER, GRASS SNAKE AND SMOOTH SNAKE.

The Adder/Viper

For the sake of clarity I will refer throughout to this reptile by its English - rather than Latin - name. Indeed up until the 1970s at least, country folk considered the `Viper` and `Adder` to be two distinct species. This belief merits further study. Apart from the normally coloured adders there are accounts of red, black, bluish-grey, whitish, red and copper and yellow adders. Of the last colour, I have only one reference. Black adders have been seen in a number of different parts of the country. They occur from the Dorset Heathlands near Studland and Poole, (see below) and they were certainly found in the New Forest in the 1930s as were the yellow variety. (They have also been known to turn up in Scotland according to Jim Foster of Herptofauna Conservation International Ltd, H.C.I.L, of Halesworth, Sussex. [2]) These were allegedly taken to the outdoor serpentarium of London Zoo in that decade and so were black ones. Black adders have been seen (though as far as I am aware not at the time of writing this, i.e. 1996) at Grovely Wood in Wiltshire, according to an eye witness, Mr. Michael Stacey of the Chalke Valley, Wilts, who once saw one there [3]. This wood is locally famous for its association with an ancient rite which allowed local inhabitants to collect as much firewood as they could from the Wood. This once led to clashes with the local landowners.

According to *The Times* of April 24th, 1937, "Among a collection of 30 adders caught in the New Forest to stock the outdoor serpentarium is an interesting example of melanism. The snake is completely black without any suggestion of pattern. Adders are very variable in coloration: the same collection includes red and yellow individuals as well as the usual brownish grey ones but these black specimens are the rarest of all. The melanism is associated with a complete absence of yellow pigment throughout the body, so that even the venom they secrete is water clear instead of straw coloured. This and other abnormal specimens are exhibited separately in the reptile house." [4].

The conventional wisdom on British snake coloration and species types is radically challenged by G.V. Wills in his "Dorset Reptiles" in 'The Dorset Yearbook 1975-1976.' [5] It is worth quoting extensively from this as the information it contains has not been found elsewhere and it shows that rustic beliefs have survived well into the age of industrialisation.

This author also believed in a separate species of British water snake distinct from any other. Hardwicke's Science Gossip of 1875 p.93 reported from Kingston of which there are several in Britain including two in Dorset: "People in this district talk much of the existence of a water-snake, which they describe as being very different from the common species, and almost black in colour. I have seen the common snake swim across a river with the grace and agility of an eel, and this latter I suspect to be the water snake of the rustics." In the Fens of the Middle Ages there was also a belief in a water snake, though this was another name for the lamprey. At the same time and place there was a belief in "large water wolves". Could this have been a species of seal? See The Medieval Fenland by H.C. Darby (p.28). Quite serious attempts by the author to track down Mr. Wills proved unsuccessful.

Wills says, "Nearly all adders are black, they only appear greyish before they are about to cast their skin, which is once a year. So as the new skin gets older it appears more grey..... There are *various species* [my emphasis] of grass snakes - some brown, some grey-brown , and those of beauty in colour with green gold and black markings. The colour photos on British wildlife in relevant books tend to be of grass snakes with the green gold and black markings". [6]

The picture below is from Thomas Bell's History of British Serpents 1839.

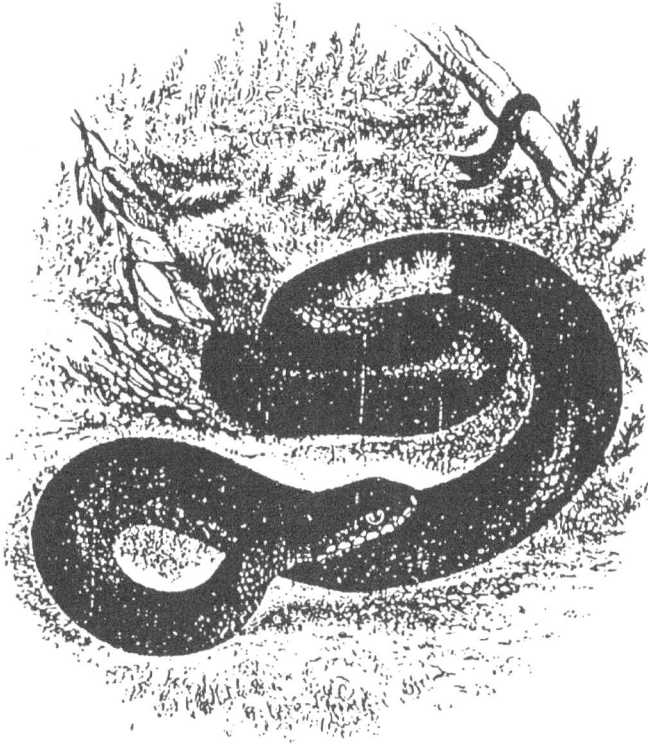

Black Adder

A more recent sighting of a black adder came to light in April 1996 on heath land in Dorset. It was discovered by two members of the Young Herpetological Club. The exact location of the find has been kept secret to protect the reptiles. I have a report (from Jon Downes of the Centre for Fortean Zoology, Summer 1996) of red adders from Leyton Bantham and Bigbury Bay in South Devon which are apparently rare because they are preyed upon by birds. Jim Foster of H.C.I.L. says, "I am not sure about red adders being female - probably, since females do tend to be more reddish, *but the coloration of juveniles and sexing them is little reported.*" (My italics.)

The article below is from 'The Hampshire Antiquary and Naturalist' Vol 1 1891 uncovered in the

Folklore Society Library London, which this author thoroughly recommends for researchers:

CURIOUS VIPER FOUND AT BITTERNE.

" E. J. M.," of Bitterne, writes under date of the 19th inst. : A day or two ago I captured a viper here with somewhat peculiar markings. At the back of the head there is a divided band of bright yellow, and immediately behind this another band of intense black. I have seen many vipers, but never one marked thus ; it is about 2ft. long. Can you inform me if it is any particular kind? Not having any spirits of wine at hand, I placed the viper in a jar of paraffin. Will it keep in this? I enclose sketch of the markings on viper's head. Replying to the enquiry, the Editor of the *Field* says : Paraffin will not keep the specimen. Vipers vary greatly in colour and markings, but we have not seen any like the one described.

The grass or ring snake(to give it one of its other names) has a yellow collar or ring behind its head or a band of yellow against black. But this is hardly a good candidate for the anomalous viper of Bitterne, as the editor of The Field, of all journals, would hardly mistake a grass snake for an adder.

In 1934 an albino adder was seen at Hembury Fort, S.E. Devon, and Thomas Bell in his `The History of British Serpents, 1839, refers to a dirty white adder with jet black markings. So further research might uncover more white adders in the South of England.

In 1887 an unnamed newspaper in the Museum at Devizes, Wiltshire reported that the then famous New Forest snake captor "Brusher Mills" asserted that the red adders never grew any larger than their usual small size and were equally as venomous as the larger brown species. Brusher Mills sold snakes to London Zoo at 1 shilling per head for the snake eating reptiles there. The largest snakes he had caught were 6ft 4 inches long and about 5 ft 2 inches. The Appendix shows a picture of a French snake capturer from 1900 with some information about him. G.R.Leighton in his `Life History of British Serpents and their local distribution in the British Isles` (1901) devotes the whole of chapter 15 to `The small red viper` in which, he states categorically, based on a survey from various parts of the country, "Careful study of British adders has driven me to regard the small red viper as a valid species, quite as distinct from the ordinary adder as a swallow is from a martin or a stoat from a weasel." [7]

Mr Leighton continues by discussing the then view, still held today, that all red adders are young female adders: "The question therefore arises, are all these so-called small red vipers simply young female adders? This could only be settled by obtaining specimens of this colour of both sexes. I had long believed in the distinctness of the small red viper from the adder, from its constancy in size and colour, but had never taken a male specimen until the 26th April 1901." [9]

This male one was found in central Dorset (Buckland Newton) where the ordinary viper was well known locally and the male there was `pale-grey with very black markings and bluish-black belly, not in the least like the small red viper`.

The picture below is from Bell's History of British Serpents and shows a red adder from Fordingbridge, Hants.

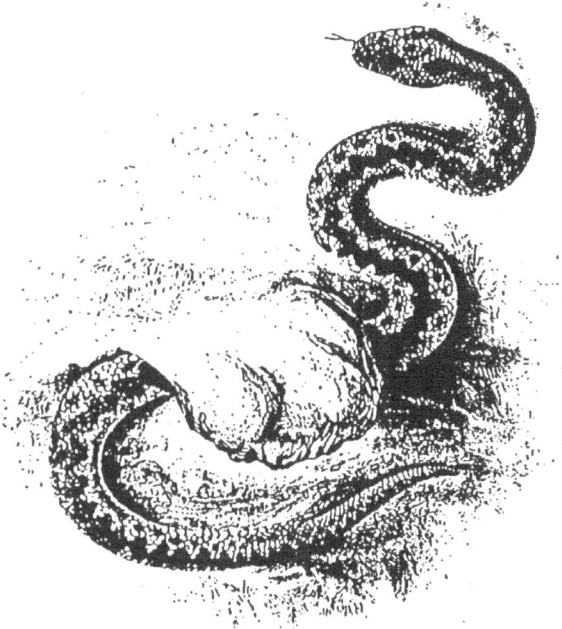

The red viper was thought to be distributed mainly in Herts, Somerset and Devon and an area near Hastings. A few authorities, other than Leighton, mention the country belief that the small red viper is more venomous and less timid than the more common adder. But in those days it was rare, so now with the destruction of their habitat it must be rarer still. But records from the Biological Records Centre in Devizes Wilts show that it has occasionally been seen in Wilts this century. Roy Pitman, a Wiltshire naturalist who found a dead polecat in Salisbury in 1986 many years after their supposed extinction in the county, provides the evidence in 1936 or 1937:

Adder: Common and widespread in suitable woodlands although it does not seem so common as formerly. Many varieties occur from black and white chequered specimens to all brown or red. Notice should be taken of a small red adder considered by some as a separate species. The colour and size being always constant and its unpleasant habit of being always so aggressive distinguishes at once this small adder from the commoner type. Like the common lizard, adders are more frequent in the S. and S.W. of the county. [9]

A Salisbury and District Natural History Society report in 1976 recorded the "rarer small variety" on Dorset heaths. Other significant records are: mention of dirty pink and orange colours and Salisbury and District Field Club Report 1955 and 1956 reported: A fine bluish-white chequered male at Redlynch 4th May 1956. The same Field Club reported in 1960: "A fine copper- coloured male habitually basked on a fallen tree trunk in Earldoms during the few spells of autumn sunshine."

The Grass Snake.

The engraving below and the accompanying article first appeared in the `Proceedings of the Zoological Society of London 1926` pp.1095-1096

P. Z. S. 1926. PROCTER. Pl. I.

John Bale Sons & Danielsson Ltd.

YOUNG ALBINO GRASS-SNAKE (Natrix natrix). Nat. size

55. A Note on an Albino Grass-Snake. By JOAN B. PROCTER, F.Z.S., F.L.S., Curator of Reptiles.

[Received October 11, 1926: Read October 19, 1926.]

(Plate I.)

Albinism in Mammals and Birds is comparatively common, but in Reptiles it has only been recorded in a few isolated instances. In the case of the Ophidia, Mr. Boulenger has observed that "partial albinism is rare ; perfect albinism, characterized by absence of black pigment in the eye, rarer still." * In certain species a pallid ivory tint is normal. In the Tree-Boa *Chondropython viridis* the young snake may be very pale yellow or pink, but is most frequently cream-colour. The green colour subsequently appears surrounding white spots, and ultimately spreads until green is the ground-colour in the adult Boa. In this case, therefore, pigment develops late in the life-history. Many desert species are pale silver-sand-colour, such as *Cerastes vipera*, *C. cornutus*, and *Crotalus mitchelli* which is sometimes called the "White Rattlesnake," but no cases of albinism have been recorded in these species.

True albinos have been known in the Grass-Snake, *Natrix natrix*, but the young specimen at present living in the Reptile House is such a perfect example, that the Garden Committee of the Society commissioned Mr. Green to make a water-colour drawing of it in order to have a permanent record of its tints in life, which will of course be destroyed by death. (Plate I.)

For the gift of this little snake we are indebted to Mrs. Durtnell, who, although thinking that it might be a viper, caught it uninjured and brought it to the Gardens in perfect condition. Its ground-colour is ivory-white, with a slightly creamier tint on the præfrontals, internasals, and first few labials. This same tint is repeated in the collar where, in a normal specimen, the yellow patches would occur. Wherever there should be black pigment, that is to say the collar, lateral spots, and ventral checkerings, the skin is as transparent as gauze, showing the colour of the muscular coat through. This is the palest mauve-pink, and produces a very odd effect, especially on the ventral surface where the liver, and other organs, show through the pale semi-transparent flesh. The characteristic collar on this white snake is therefore shown up in cream and pale mauve-pink.

The eyes have brilliant dark red pupils, surrounded by a more opaque pale orange iris, and entirely lack that dull, blind look seen in pink-eyed mammals. The nostrils show up their pink linings, and the tongue is pink at the base and ivory

* Boulenger, G. A. 'The Snakes of Europe,' 1913, p. 39 (Methuen, London).

on the bifurcated portion. The general effect is very pale and fragile like carved ivory, with jewel eyes.

This is the second albino snake which has been on view in the Reptile House. The first, an Indian Cobra, *Naia tripudians*, was a true albino in the black pigment sense, but had a powdering of sand-coloured chromatophores on the hood, forming shadowy white "spectacles" on a sandy ground-colour. An account of this snake, together with a photographic plate, appeared in the Proceedings in 1924 *, and it lived in perfect health for 22 months. During this time we were fortunate enough to obtain an excellent cinematograph record of it, taken out of doors in natural surroundings, and a copy of this film is preserved in our library.

Most of the albino snakes have this trace of sandy or cinnamon coloured pigment emphasising the markings characteristic of the species. Mr. Walter Goodfellow has seen a white Reticulated Python with faint markings of this kind, and with red eyes, in Singapore. Other records are as follows :—Ditmars †, in 20 years, has seen an albino Palm Viper, a Rattlesnake (species not indicated), a Milk Snake, and a Black Snake. Klauber ‡ has published a photograph of an albino Gopher Snake, and Boulenger §, in 'Snakes of Europe,' mentions the Grass-Snake, Tesselated Snake, Æsculapian Snake, and Smooth Snake.

* Procter, J. B., P. Z. S. 1924, p. 1125, pl. i.
† Ditmars, Zool. Soc. Bull., New York, xxiv. no. 6, 1924, p. 127.
‡ Klauber, Zool. Soc. Bull., New York, xxvii. no. 3, 1924, p. 80.
§ Boulenger, 'Snakes of Europe,' 1913, p. 39.

The Smooth Snake

As of the time of writing, mid 1997, no colour variations of the smooth snake have been found as far as I am aware. According to a spokesman for a Hampshire Herpetological Society none had been recorded to the best of his knowledge.

To conclude, I include some mystery snakes from Scotland. They were "found" in a reproduction of a late 17th Century travel book to the Western Islands of Scotland: `A Description of The Western Islands of Scotland Circa 1695` by M. Martin. Reprinted in 1934 and 1994.

The same work also mentions the King Otter which was also believed in in Western Ireland." Serpents abound in several parts of this isle; there are three kinds of them, the first black and white spotted, which is the most poisonous, and if a speedy remedy be not made use of after the wound given, the party is in danger. I had an account that a man at Glenmore, a boy at Portree, and a woman at Loch-Scahvag, did all die of wounds given by this sort of serpents...... The longest of the black serpents mentioned above is from two to three, or at most four feet long. The yellow serpent with brown spots, is not so poisonous, nor so long as the black and white one. The brown serpent is of all three the least poisonous and smallest and shortest in size. [11] "The "isle" referred to above is The Isle of Skye.

British and Chinese snake catchers.

The following information is taken from `The Country-Side` magazine of August 5th 1905 (p.199) and amounts to a belated obituary to one of my favourite figures loosely related to exotic zoology. Brusher Mills ought to be rescued from his undeserved neglect of almost a century. The article is not reproduced at length but the interesting pieces are mentioned, as follows:

"Not with the magic flute of Pan, or Orpheus` lyre does he lure the snakes from their thickets, but pursues his somewhat dangerous craft armed only with a stout hazel stick with a forked tip, and a large pair of iron pincers like the inquisitors of old..... His largest catch was 160 adders in a month, and the average bag in two years was five thousand serpents, of which one thousand were adders... He is a handsome old man, looking rather like Rip van Winkle tidied up, with his long beard and thick eyebrows.

"He is to be met with in all weathers, tramping the Forest roads or emerging from one of the enclosures, to all of which he has free access. He is hung all over with sacks and bags, and he carries besides his stick and pincers a big tin which holds his quarry.... Brusher is a firm believer of the old tradition that adders fat is a sovereign remedy for bruises, black eyes, rheumatic gout, and a host of other ills.... Salt butter is the only cure for a snake bite, so he says. Hold up your hand, or whatever member has been bitten, let the blood flow well, and then rub in a good quantity of salt butter, which is a certain cure... He manages to make a living by boiling down the adders` fat and selling it, besides selling the finer specimens alive."

The following extract is from the Hong Kong *South China Morning Post* of October 28th 1933 (p.2):

"We have some remarkably efficient snake catchers in this Colony, and they have their counterparts in various districts throughout China. I have watched them at work - I once saw a man seize a cobra about five feet in length and whisk it into a bag before the reptile had time to realise what was happening. There are cases on record locally of these men being bitten by venomous snakes and recovering, under their own treatment. Questions of immunity, induced or inherited, are subjects for medical discussion: it is certain that the professional Chinese snake catcher can take risks which to the ordinary human being would be fatal."

The paper then goes on to quote from "a recent issue" of the North China Daily News: "In your issue of 25th. September you gave an account of a gentleman by the name of Mr Wang Hai-san, a native of Shanghai, who was in Chefoo catching snakes by rubbing spice on the end of his finger, and allowing them to bite. It may interest you to know that about two years ago I saw a man, perhaps Mr. Wang, doing this only a few miles from Shanghai; he was carrying a basket containing about forty snakes, apparently caught that day. I watched him put his finger into the few inches of space under a coffin, he made a clucking noise and within a few seconds slowly pulled out a snake which had its fangs deeply embedded in the end of his fingers..."

More British snake anomalies.

Picking up books at random off the shelves of book shops can be as effective a way of finding interesting cryptozoological material as working in a reference library. In Shaftesbury a while ago I was doing just the former when I came across an un-snakelike book entitled Wessex Country Tales - the Memories of A Farmer's Boy, by Bernard Dunford. In it he gives an account of an anomalous snake seen somewhere in S. Dorset. (Probably near Dorchester in the early 1920s, though the text is unclear. There are several Dunfords in a contemporary Dorset trade directory and attempts to contact Mr Dunford a year or so ago were not successful.):

"We moved in March. It was a cold spring and the snakes were still hibernating but, as the weather became warmer, they began to appear. The land behind the house had not been cleared, and was merely used for grazing sheep. There was an abundance of gorse and other shrubs, so it was ideal for playing hide and seek. I found a clump of brambles, crouched down and kept as still as possible.... I looked down and saw a snake crawl between my legs... When I told people about it they invariably asked what kind of snake it was; it was a light brownish colour with a black zig-zag stripe on its back and looked about six feet long and six inches round. I was told there are no snakes as big as that in the British Isles, but what do they know about it? It was my snake. I saw it and they didn't." [12]

In the 1980s, garter snakes were turning up in Dorset. According to the Dorset Echo of Oct 11th 1984 a specimen of the North American garter snake was found by a 'Weymouth man' but the paper didn't say where. But the Dorset Environmental Records Centre should know. In July 1953 according to the British Journal of Herpetology v.1(9) 1953 p.174, two Viperine snakes, Natrix maura, turned up in Sidcup, Kent. A notice in the Sidcup and Kentish Times failed to trace their origin.

More recently in the early 1960s a single gravid female aesculapean snake escaped from Welsh Mountain Zoo at Colwyn Bay in N. Wales. By September 1984 "the snakes [had] been seen up to one km from the zoo," (B.B.C Wildlife Magazine Sept.1984 p.430) but a conversation with a spokesperson at the zoo in 1996 did not confirm this. There is an excellent colour photo of this snake in *The Times* of July 21st 1997, p.7.

In 1987 a 6ft long snake, possibly a boa, turned up in straw bales on a farm near Newquay. According to the farmer on whose land it was found it was preying on lambs and might have escaped from a neighbour. According to the *Daily Mirror* of June 30th 1987, "The poisonous giant has already killed two sheep and injured a cow."

Earlier in the year, ironically also from Newquay, Mr Rose was driving along the M5 when an adder poked its head out of the ventilator. Mr Rose had hired the car from Wadebridge in Cornwall. There was little the driver could do as the car was in a contraflow system. "A spokesman for the firm said: `We`ve got no idea where the snake came from." (*Daily Mail* April 25th 1987)

According to the *News of The World* for August 10th 1997 a fisherman caught an adder a mile off Lyme Regis, Dorset. It is recorded in English Medieval Graffiti (V.Pritchard,Cambridge University Press 1967 p.167) that on the south-east pier of the central tower of Wheathampstead Church, Herts, there is a graffito of a snake with a single bent arm-like leg. This is from about the Thirteenth Century.

According to an unnamed Northamptonshire newspaper of March 1760 (possibly the *Northampton Mercury*): "Last Monday as a man was digging up the root of an old tree in a wood near Northampton, e was surprised by the hissing of snakes or adders, whereupon he called to his companion who was near at hand, in order to make a further search; when, to their great surprise, they discovered near 100 snakes of diverse colours and a bed of birds wings, and feathers, among which they secured a serpent nine inches round the middle, and five feet in length, which was shewn to a great number of spectators."

Going forward in time, *The Times* of September 19th 1788 included the following strange story from Hodsdon (i.e. Hoddesdon) also in Hertfordshire. "On Thursday last, as a farmer belonging to Hodsdon in Hertfordshire, was travelling his grounds with his gun, he observed an uncommon rustling in some branches- curiosity induced him to advance, imagining that a hare was intangled in the branches, but what was his astonishment on his arrival at the place , to behold an enormous snake- which with erected crest and dreadful hissings- threatened him" The piece goes on to state that the farmer fled but later returned with a friend and shot the snake. On being stretched out on the path "it measured 12 feet from the head to the extreme part of the tail and the circumference in the thickest part 14 inches."

This was not the last in a line of large British snakes. According to Notes and Queries for Naturalists for August 1st 1857 (p.72) a giant snake turned up near Colchester sometime in 1855; exact date unknown:

A MONSTER ENGLISH SNAKE -

The most extraordinary example of the snake species ever discovered in this country was found dead on a farm in the suburbs of Colchester, last week. It was 9 feet 5 inches long, 11 inches in girth at the thickest part, and was thought to weigh 14lb or 15lb."

Actually the Herts snake is more extraordinary than this. Both locations are quite far from the sea so the hypothesis that they arrived by ship is not very plausible. Around about 1910 there were interesting beliefs in the area of the village of Nomansland, on the Wiltshire-Hampshire border near the New Forest, with regard to snakes. (There is a pub near this village which about a year ago had in it a stuffed dog-like fox or fox-like dog!) According to a booklet Nomansland A Village History: "To begin with, then, a clear distinction is drawn between the adder and the viper." [13]

"The Viper, Length about 7 and a half inches, colour buff with dark brown markings and a V on the head.... It will not try to get away, and is always ready for action." [14]

"The Adder. The length of the female is 16 or 17 inches, the male being two to three inches less. Colour buff ,with dark brown markings, or brown, with black markings, or black, with white markings..... The offspring, it is said, are always an odd number: 7, 9, 11 or 13... The adder has an oil gland in the neck which acquires a full charge of oil by March. "Adders fat" is a valuable antidote to the creatures own poison, and a useful remedy for the stings of bees and flies. The adder and viper are both viviparous." [15]

"It is affirmed throughout the Forest that there is a casual hybrid between the (Grass) snake and the adder, which in consequence of being neither adder or snake is known as a Neither (pronounced nither) ... In proportions it runs closer to the adder than to the snake, being about 18 inches long; that is, rather longer than an adder but not quite so stout." [16]

Another interesting British snake story is from the *Diary of Samuel Pepys*, 4th February 1662. In it the famous diarist wrote:

"To Westminster-hall, where it was full terme. Here all the morning; and at noon to my Lord Crewes - where one Mr. Templer (an ingenious man and a person of honour he seems to be) dined; and discoursing of the nature of Serpents, he told us of some that in the waste places in Lancashire do grow to a great bigness, and do feed upon larkes, which they take thus - they observe when the lark is soared to the highest, and do crawle till they come to be just underneath them; and there they place themselves with their mouth uppermost, and there (as is conceived) they do eject poyson up to the bird; for the bird doth suddenly come down again in its course of a circle, and falls directly into the mouth of the serpent - which is very strange." [17]. (Thanks to Mrs A. Fitzsimons, Assistant Librarian, Pepys Collection, Magdalene College, Cambridge for this information.)

Interestingly, according to a Salisbury and District Natural History Society report for 1962 in November of that year a 3ft long female grass snake caught by a boy near High Post disgorged a fully grown sky lark!

There is a legend, recorded in Legends and Traditions of Yorkshire (1889, author unknown, p.238) of a battle at Kellington, near Pontefract between a shepherd, Armroyd and his dog and a giant serpent. This battle in which shepherd, dog and snake were all killed, took place some time in the dim and distant past. There was, in the 1880s a stone which faintly portrayed this battle in the churchyard of Kellington.

The Giant Snakes of the Ukraine and Ireland.

In the early 19th Century there were German settlements in what is now the Crimea and southern Ukraine. In the early 1840s German settlers and Russians lived in close proximity on the steppe and reed grounds of the Dniester River. An eighteen inch long species of lizard lived here as did the snake coluber trabalis which could grow up to eighteen feet long but specimens had been recorded at twenty two and a half feet. "Legends are not wanting among the Cossacks of gigantic serpents that, at no very remote period, infested the reed grounds of the Dniester, whence they sallied forth to kill men and oxen, and now and then to amuse themselves by running down a rider and his steed, no horse being fleet enough to effect its escape..." [18]

On another occasion a giant snake frightened some German agricultural labourers out on the steppe with their horses. Large tracks were found through the cornfields the width of a sack of flour. Some shepherds on encountering the snake, "...fled with their flocks in dismay, but not before the huge reptile had killed one of their horses before their faces." [19] A team of about one hundred young men went out to find and kill the snake and on sighting it they shot several times wounding it and it took flight till lost in the Dniester reed beds. "Some of the more imaginative among the sportsmen insisted upon it that the snake was at least thirty feet long. The Schulze, whose computation was the most moderate, and probably the nearest to truth, calculated the length of the animal to be at least three and a half fathoms." The Schulze was the local magistrate and three and a half fathoms would be 21 feet. According to correspondence with Chris Moiser, this "snake" could be a type of giant cat fish.

The Russians were not willing at this time to pursue and kill snakes. This was because they believed that if a snake was killed a corporate act of revenge on the snake killer would spread to all the other snakes in the area and the human snake killer would himself end up dead. This belief was based on Acts of the Apostles (Bible) ch.28 verses 3-6 when Paul picks up a poisonous snake without being harmed and at first the Maltese think he is a murderer; then when he remained alive they thought he was a god. The Russian interpretation of this event was that snakes take vengeance on murderers but especially snake murderers.

Most curiously, like a good deal of folklore, there is a parallel elsewhere in the world. In Warwickshire - specifically, Alcester - there occurred the following: "On December 16th, 1544 in the town of Alcester appeared on this day a great Adder which did so bite a man that he swelled so large that it required a coffin three times the usual size in which to bury him, which was done the next day at sunset. Legend says that this adder had specially journeyed to Alcester to punish this man for certain sins he had committed and it was never seen again." [20] Conversations with Darren Naish and Chris Moiser in the summer of 1997 failed to establish the identity of these Crimean snakes but the latter suggested they may really have been a species of European catfish.

There are occasional reports of snakes turning up in Ireland, though none recently. The report below is from The Wexford Independent of 29th June 1901 p.2 (I have others on file) :

"Sir - In reference to the snake caught by me on last Sunday week, I wish to state that it is alive and well and has been given to the man P. Myrtle whose dog discovered it. It is believed the snake came over from England in some shrubs which were imported last year. I beg to say there is no foundation for the statement about 'a lot of money being got on the snake.' There was no charge for seeing the snake."

In 1888 a series of thefts or the devil was thought to be the reason for the disappearance of sheep pigs and poultry in the vicinity of Amraugh and Castleraine in Ireland towns twelve miles apart. After prayers failed, two detectives were called from Dublin who actually witnessed snakes killing and carrying off the prey. They were 15ft long and of a dark colour. The Dublin Freeman's Journal then recalled that an American showman had got drunk one night in Amraugh in 1885 and released his menagerie which included snakes which had multiplied. "And in fact they kept cropping up in various parts of Ireland at uncertain intervals, and a militant union of Church and State was found necessary to suppress them entirely." [21]

Notes and References

1 Leighton G, *Life of British Serpents* 1901, chapter 15.

2 Letter from Jim Foster of H.C.I.L. 1996

3 Conversation with Michael Stacey, summer 1996

4 Anon: "Brazilian Snakes for the Zoo, New Forest Adders" in *The Times* 24 Apr 1937.

5 Wills G V, "Dorset Reptiles" in *The Dorset Yearbook 1975-1976*, pp128-129.

6 Ibid.

7 Foster op. cit.

8 Leighton G, op. cit. p206

9 Leighton G, op. cit. p209

10 From cards at The Biological Records Centre, Devizes, Wiltshire, UK.

11 Ibid: M Martin "A description of the Western Isles of Scotland including a Voyage to St Kilda".

12 Dunford B "Wessex Country Tales: the Memories of a Farmer's Boy" (Whittlebury: Sporting & Leisure Press 1992), p19.

13 Livens H M, "Nomansland: A Village History" (Salisbury Times & South Wilts Gazette, 1910) p43.

14 Ibid p44

15 Ibid pp44-45

16 Ibid p45

17 Latham and Matthews ed, vol 3, of *The Diaries of Samuel Pepys* (Harper Collins), p22.

18 Author unknown: "The Steppes of Southern Russia - No.2" in *The Asiatic Journal and Monthly Register for British and Foreign India, China and Australasia* vol 36 - new series (Sep-Dec 1841) p223.

19 Ibid p224

20 Gwinnett A J, *A History of Alcester* p94

21 Walsh W S, *A Handbook of Curious Information* (J B Lippincott, London) p812.

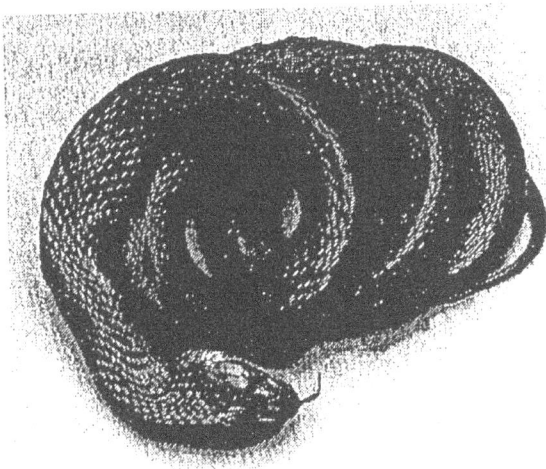

1997 - A Year in the Life of
The Centre for Fortean Zoology

by

Jonathan Downes

This was the year (at the risk of sounding like the introduction to the Channel 4 TV series *Babylon 5*) that it all happened. After four years languishing in the metaphorical wilderness of self-publication and outdated equipment, we finally managed not only to obtain some proper equipment for a change but to make impressive forays into the world of `proper` publishing with a monthly column in `Uri Geller's Encounters` and regular features in a number of other magazines. We also managed to secure a regular fortnightly radio series on BBC Radio Devon and at the time of typing this are just about to fly to the U.S., Mexico and Puerto Rico courtesy of Channel 4 Television to film the first proper CFZ expedition in search of the semi-mythical chupacabras - the blood sucking vampiric zooform entity of hispanic Central America.

On the down-side, the massive public interest in things fortean appears to be on the wane and although 1997 was a year in which we managed to make a good deal of media headway, it seems unlikely that this will continue. Several of the newsstand magazines for which we had been writing regularly have now ceased publication and it seems likely that others will go within the next twelve months. As a result of this we will have to find alternate forms of funding because as we have found out over the past four years running the Centre for Fortean Zoology is a full time job in itself, and we would not be able to do what we do if we had to have "proper" jobs as well!

We have now, for the first time managed to acquire computer equipment which is not hopelessly outdated and during the final weeks of 1997 we made our first tentative forays into cyberspace. For this we have to proffer our undying thanks to Dave Simons from Derby, an old friend from my mis-spent youth who set us up with our first web-site and offered invaluable advice and help as we set up the second and far more complex version which is now available at www.cclipsc.co.uk/cfz. The Internet is, like it or not (and personally we LOVE it) the future, and there is no doubt that whatever we do next, our adventures in cyberspace will be as important as our adventures in foreign lands.

Looking forward to 1998, our biggest and most important New Year's Resolution is that we continue to improve our service and that with the advent of our new computer system both Animals & Men and our other publications shall go from strength to strength. We intend to relaunch Animals & Men with a radical new image in the spring of this year and we hope that the typographic and design changes that we are planning will be generally acceptable.

Because of this, and also because of our Central American excursions we are, for the first time since its inception four years ago, `missing` an issue of the magazine. The issue which was scheduled to appear in January 1998 is not being published but from April the normal service will be resumed.

It is too early to say what the new year will bring. It is full of promise and we have several exciting projects that must (at the moment at least) remain under wraps.

During 1997 we managed to publish two books and launch a second magazine and a web-site. During 1998 we sincerely hope that that this progress will continue.

Thank you for your support during the last twelve months. here's to another year of high strangeness within the animal kingdom.

Best wishes,

Jon Downes
(Director of the Centre for Fortean Zoology).

CFZ WEBSITE

The CFZ is now on-line! We can be found at:

http://www.eclipse.co.uk/cfz

THE CENTRE FOR FORTEAN ZOOLOGY

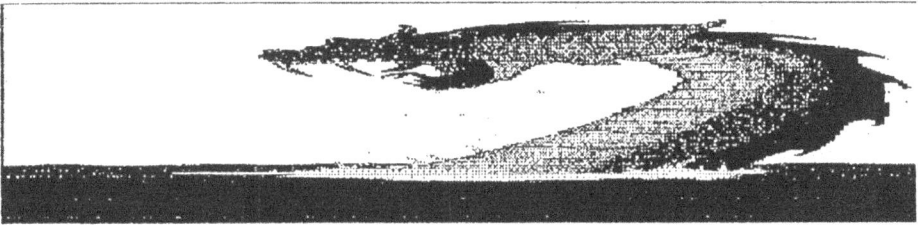

OPTIONS - please scroll down and click on one of the underlined items:

Explore <u>this CFZ site</u> further, for news, features and information...

Or visit our sister sites:

<u>The Goblin Universe</u> - The parish magazine of the outer edge: the world of ghosts UFOs and the paranormal

The <u>Exeter Strange Phenomena</u> (ESP) Research Group site, researching events such as UFO clusters

Our website is ever-expanding: by the time you read this, more material will have been added.

From the initial page, the **main CFZ pages** can be accessed:

There are features on mystery insects, the Owlman, mystery animals of Hong Kong, BHMs, and Morgawr. There's also a description of our Latin American hunt for the Chupacabra.

Additional pages include a guide to our publications - "Animals & Men", etc, and a page of addresses / links to other web sites. Other pages will be added when needed

Our eMail address is **jon@eclipse.co.uk**

The following can also be accessed from the initial page:

The Goblin Universe

Weird stuff, UFOs, general forteana. The "Parish Magazine of the Outer Edge".

The Exeter Strange Phenomena (ESP) Research Group pages

Local investigations of anything unexplained, including some cryptozoology, and Jonathan's tour guide, "Weird About The West". There's also a page covering Jonathan and Graham's radio show on BBC Radio Devon. Past guests have included Tony Shiels, Richard Freeman, Jenny Randles, Darren Naish, Lionel Fanthorpe, Chris Moiser, Sally Parsons and Dr Karl Shuker.

THE CENTRE FOR FORTEAN ZOOLOGY

So, what is the Centre for Fortean Zoology?

We are a non profit-making organisation founded in 1992 with the aim of being a clearing house for information, and coordinating research into mystery animals around the world. We also study out of place animals, rare and aberrant animal behaviour, and Zooform Phenomena; little-understood "things" that appear to be animals, but which are in fact nothing of the sort, and not even alive (at least in the way we understand the term).

Why should I join the Centre for Fortean Zoology?

Not only are we the biggest organisation of our type in the world, but - or so we like to think - we are the best. We are certainly the only truly global Cryptozoological research organisation, and we carry out our investigations using a strictly scientific set of guidelines. We are expanding all the time and looking to recruit new members to help us in our research into mysterious animals and strange creatures across the globe. Why should you join us? Because, if you are genuinely interested in trying to solve the last great mysteries of Mother Nature, there is nobody better than us with whom to do it.

What do I get if I join the Centre for Fortean Zoology?

For £12 a year, you get a four-issue subscription to our journal *Animals & Men*. Each issue contains 60 pages packed with news, articles, letters, research papers, field reports, and even a gossip column! The magazine is A5 in format with a full colour cover. You also have access to one of the world's largest collections of resource material dealing with cryptozoology and allied disciplines, and people from the CFZ membership regularly take part in fieldwork and expeditions around the world.

How is the Centre for Fortean Zoology organized?

The CFZ is managed by a three-man board of trustees, with a non-profit making trust registered with HM Government Stamp Office. The board of trustees is supported by a Permanent Directorate of full and part-time staff, and advised by a Consultancy Board of specialists - many of whom who are world-renowned experts in their particular field.   We have regional representatives across the UK, the USA, and many other parts of the world, and are affiliated with other organisations whose aims and protocols mirror our own.

I am new to the subject, and although I am interested I have little practical knowledge. I don't want to feel out of my depth. What should I do?

Don't worry. We were *all* beginners once. You'll find that the people at the CFZ are friendly and approachable. We have a thriving forum on the website which is the hub of an ever-growing electronic community. You will soon find your feet. Many members of the CFZ Permanent Directorate started off as ordinary members, and now work full-time chasing monsters around the world.

I have an idea for a project which isn't on your website. What do I do?

Write to us, e-mail us, or telephone us. The list of future projects on the website is not exhaustive. If you have a good idea for an investigation, please tell us. We may well be able to help.

How do I go on an expedition?

We are always looking for volunteers to join us. If you see a project that interests you, do not hesitate to get in touch with us. Under certain circumstances we can help provide funding for your trip. If you look on the future projects section of the website, you can see some of the projects that we have pencilled in for the next few years.

In 2003 and 2004 we sent three-man expeditions to Sumatra looking for Orang-Pendek - a semi-legendary bipedal ape. The same three went to Mongolia in 2005. All three members started off merely subscribers to the CFZ magazine.

Next time it could be you!

Project Kerinci, Sumatra - 2003
In search of the bipedal ape Orang Pendek

How is the Centre for Fortean Zoology funded?

We have no magic sources of income. All our funds come from donations, membership fees, works that we do for TV, radio or magazines, and sales of our publications and merchandise. We are always looking for corporate sponsorship, and other sources of revenue. If you have any ideas for fund-raising please let us know. However, unlike other cryptozoological organisations in the past, we do not live in an intellectual ivory tower. We are not afraid to get our hands dirty, and furthermore we are not one of those organisations where the membership have to raise money so that a privileged few can go on expensive foreign trips. Our research teams both in the UK and abroad, consist of a mixture of experienced and inexperienced personnel. We are truly a community, and work on the premise that the benefits of CFZ membership are open to all.

What do you do with the data you gather from your investigations and expeditions?

Reports of our investigations are published on our website as soon as they are available. Preliminary reports are posted within days of the project finishing.

Each year we publish a 200 page yearbook containing research papers and expedition reports too long to be printed in the journal. We freely circulate our information to anybody who asks for it.

No. Each year since 2000 we have held our annual convention - the *Weird Weekend* - in Exeter. It is three days of lectures, workshops, and excursions. But most importantly it is a chance for members of the CFZ to meet each other, and to talk with the members of the permanent directorate in a relaxed and informal setting and preferably with a pint of beer in one hand. Since 2006 - the *Weird Weekend* has been bigger and better and held in the idyllic rural location of Woolsery in North Devon. The 2008 event will be held over the weekend 15-17 August.

Since relocating to North Devon in 2005 we have become ever more closely involved with other community organisations, and we hope that this trend will continue. We also work closely with Police Forces across the UK as consultants for animal mutilation cases, and we intend to forge closer links with the coastguard and other community services. We want to work closely with those who regularly travel into the Bristol Channel, so that if the recent trend of exotic animal visitors to our coastal waters continues, we can be out there as soon as possible.

We are building a Visitor's Centre in rural North Devon. This will not be open to the general public, but will provide a museum, a library and an educational resource for our members (currently over 400) across the globe. We are also planning a youth organisation which will involve children and young people in our activities. We work closely with *Tropiquaria* - a small zoo in north Somerset, and have several exciting conservation projects planned.

Apart from having been the only Fortean Zoological organisation in the world to have consistently published material on all aspects of the subject for over a decade, we have achieved the following concrete results:

- Disproved the myth relating to the headless so-called sea-serpent carcass of Durgan beach in Cornwall 1975
- Disproved the story of the 1988 puma skull of Lustleigh Cleave
- Carried out the only in-depth research ever into the mythos of the Cornish Owlman
- Made the first records of a tropical species of lamprey
- Made the first records of a luminous cave gnat larva in Thailand.
- Discovered a possible new species of British mammal - the beech marten.
- In 1994-6 carried out the first archival fortean zoological survey of Hong Kong.
- In the year 2000, CFZ theories where confirmed when an entirely new species of lizard was found resident in Britain.
- Identified the monster of Martin Mere in Lancashire as a giant wels catfish
- Expanded the known range of Armitage's skink in the Gambia by 80%
- Obtained photographic evidence of the remains of Europe's largest known pike
- Carried out the first ever in-depth study of the *ninki-nanka*
- Carried out the first attempt to breed Puerto Rican cave snails in captivity
- Were the first European explorers to visit the `lost valley` in Sumatra
- Published the first ever evidence for a new tribe of pygmies in Guyana
- Published the first evidence for a new species of caiman in Guyana

EXPEDITIONS & INVESTIGATIONS TO DATE INCLUDE:

- 1998 Puerto Rico, Florida, Mexico *(Chupacabras)*
- 1999 Nevada *(Bigfoot)*
- 2000 Thailand *(Giant snakes called nagas)*
- 2002 Martin Mere *(Giant catfish)*
- 2002 Cleveland *(Wallaby mutilation)*
- 2003 Bolam Lake *(BHM Reports)*
- 2003 Sumatra *(Orang Pendek)*
- 2003 Texas *(Bigfoot; giant snapping turtles)*
- 2004 Sumatra *(Orang Pendek; cigau, a sabre-toothed cat)*
- 2004 Illinois *(Black panthers; cicada swarm)*
- 2004 Texas *(Mystery blue dog)*
- 2004 Puerto Rico *(Chupacabras; carnivorous cave snails)*
- 2005 Belize *(Affiliate expedition for hairy dwarfs)*
- 2005 Mongolia *(Allghoi Khorkhoi aka Mongolian death worm)*
- 2006 Gambia *(Gambo - Gambian sea monster , Ninki Nanka and Armitage s skink*
- 2006 Llangorse Lake *(Giant pike, giant eels)*
- 2006 Windermere *(Giant eels)*
- 2007 Coniston Water *(Giant eels)*
- 2007 Guyana *(Giant anaconda, didi, water tiger)*

To apply for a **FREE** information pack about the organisation and details of how to join, plus information on current and future projects, expeditions and events.

Send a stamped and addressed envelope to:

THE CENTRE FOR FORTEAN ZOOLOG
MYRTLE COTTAGE, WOOLSERY,
BIDEFORD, NORTH DEVON
EX39 5QR.

or alternatively visit our website at:
w w w . c f z . o r g . u k

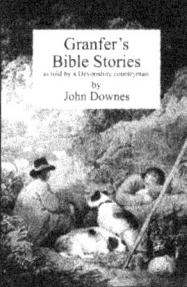

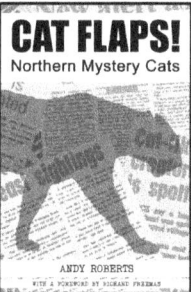

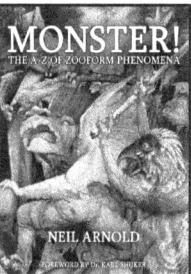

Other books available from
CFZ PRESS

Other books available from
CFZ PRESS

.

www.ingramcontent.com/pod-product-compliance
Lightning Source LLC
Chambersburg PA
CBHW072127270326
41931CB00010B/1691